Carpentry and Joinery

1

Carpentry and Joinery

1

Second Edition

Brian Porter, LCG, FIOC
Department of Building Trades, Leeds College of Building

General Technical Adviser and Consultant:
W.R. Rose MCIOB, DMS, DASTE, MASI, FIOC

Woodworking Machines:
Consultant and Contributor: Alan Wilson LCG, MIM WOOD.T
Consultant: Eric Cannell

A member of the Hodder Headline Group
LONDON • SYDNEY • AUCKLAND

First published in Great Britain in 1990 by Edward Arnold
a member of the Hodder Headline Group
338 Euston Road, London NW1 3BH

British Library Cataloguing in Publication Data
Porter, Brian 1938-
Carpentry and joinery–2nd ed.
Vol. 1
1. Carpentry and joinery.
I. Title
694

ISBN 0 340 50773 X

9 10 11 12 13 14 15

Typeset in 10/11 Baskerville by Colset Private Ltd, Singapore
Printed and bound in Great Britain by The Bath Press, Bath

Contents

Preface   vi

Acknowledgements   vii

1   Timber and associated materials   1
Growth and structure of a tree. Hardwoods and softwoods. Sources and supply of timber. Size and selection of timber. Drying timber and moisture content. Common enemies. Manufactured boards. Laminated plastics. Board and sheet sizes.

2   Hand tools and workshop practices   27
Measuring tools. Setting-out and marking-out (marking-off) tools. Saws. Planes. Boring tools. Chisels. Shaping tools. Finishing tools and abrasives. Holding equipment (tools and devices). Tool storage. Tool maintenance.

3   The Porterbox storage system   74
Practical project 1: Porterbox drop-fronted tool box and saw stool. Practical project 2: Portercaddy. Practical project 3: Portercase. Practical project 4: Porterdolly. Practical project 5: Porterchest.

4   Portable powered hand tools   88
Electric drills. Rotary impact drills. Electric screwdriver. Belt Sander. Orbital sander or finishing sander. Specification plate (SP). General safety.

5   Woodworking machines   97
Cross-cutting machines. Circular sawing machines. Planing machines. Mortising machines. Narrow band-sawing machines. Wood-turning lathes. Sanding machines. Woodworking machines regulations 1974. Grinding machines. Workshop layout.

6   Basic woodworking joints   111
Lengthening – end joints. Widening – edge joints. Framing – angle joints.

7   Suspended timber ground floors   120
Floor joists. Flooring (decking). Skirting.

8   Gable-ended single roofs   127
Roof terminology. Forming a pitched roof. Common rafters. Roof assembly. Eaves details.

9   Turning pieces and centres up to 1 metre span   134
Simple geometric arch shapes. Turning pieces. Centres. Easing and striking centres.

10   Formwork   141
Formwork design. In *situ* work (shuttering) Pre-cast work (mould boxes).

11   Ledged-and-battened doors   145
Door construction. Assembly. Framing. Fitting and hanging. Items and processes.

12   Single-light casement windows   152
Construction. Fixing to the structure.

13   Moisture movement   157
Surface tension. Capillarity. Preventative measures.

14   Shelving   160
Traditional shelving. Proprietary systems and shelving aids.

15   Fixing devices   163
Nails. Wood screws. Threaded bolts. Fixing plates. Plugs.

Index   175

Preface

The main difference between this book – the first of three new editions – is the overall size and format of the contents compared to the previous first edition (published in 1982). The most important reason for this change is that in the majority of cases text and illustration can now share the same or adjacent page, making reference simpler and the book easier to follow. The most significant change of all is the new section on tool storage: Several practical innovative projects have been included which will allow the reader to make, either as part of his or her coursework or as a separate exercise, a simple, yet practical system of tool storage units and tool holders – aptly called the 'Porterbox' system (original designs were first published in *Woodworker* magazine). Each chapter has been reviewed and revised to suit current changes. For example, this has meant the introduction of new hand tools, replacing or supplementing existing portable powered hand tools, and updating some woodworking machines.

Educational and training establishments seem to be in a constant state of change; college and school based carpentry and joinery courses are no exception. As the time available for formal tution becomes less, course content, possibly due to demands made by industry and the introduction of new materials together with a knowledge of any associated modern technology they bring with them, seems to be getting greater – making demands for support resource material probably greater than they have ever been.

Distance learning (home study with professional support) can have a very important role to play in the learning process, and it should be pointed out that in some areas of study it is not just an alternative to the more formally structured learning process, but a proven method in its own right.

No matter which study method is chosen by the reader, the type of reading matter used to accompany studies should be easy to read and highly illustrative, and all the subjects portrayed throughout this book meet that requirement. I hope therefore that this book is as well read, and used, by students of this most fulfilling of crafts.

Brian Porter
Leeds 1989

Preface to the first edition

This volume is the first of three designed to meet the needs of students engaged on a course of study in carpentry and joinery. Together, the three volumes cover the content of the City and Guilds of London Institute craft certificate course in carpentry and joinery (course number 585).

I have adopted a predominantly pictorial approach to the subject matter and have tried to integrate the discussion of craft theory and associated subjects such as geometry and mensuration so that their interdependence is apparent throughout. However, I have not attempted to offer instruction in sketching, drawing, and perspective techniques (BS 1192), which I think are best left to the individual student's school or college.

Procedures described in the practical sections of the text have been chosen because they follow safe working principles – this is not to say that there are no suitable alternatives, simply that I favour the ones chosen.

Finally, although the main aim of the book is to supplement school or college-based work of a theoretical and practical nature, its presentation is such that it should also prove invaluable to students studying by correspondence course ('distance learning') and to mature students who in earlier years may perhaps have overlooked the all-important basic principles of our craft.

Brian Porter
Leeds 1982

Acknowledgements

It would have been almost impossible to have produced this book without the help and guidance given by friends and colleagues during the writing of the first edition, in particular Mr Edward Judkins MCIOB, AIOC, senior lecturer at Leeds College of Building. I am indebted to Mr W.R. Rose (Head of Department) for currently undertaking the role of technical editor, and consultant advisor, also to Alan Wilson LCG, MIMWood.T, for his work in updating and contributing to the chapter on woodworking machines, together with the help given by Eric Cannell, who also kindly proof-read work relating to timber technology, and to Mr R.C. Smith for contributing work on voltage operated earth leakage circuit breakers (ELCBs).

The photographs in Figs 2.15–2.17(a), 2.18–2.22, 2.40, 2.41, 2.43, 2.45–2.47, 2.51, 2.66–2.68, 2.79, 2.80, 2.82, 2.85, 2.86 and 2.98(b) were taken by Mr J.H. Dwight and Mr K. Procter, and I am very grateful for their help.

I would also like to thank Mr P. Peck (Record Marples (Woodworking Tools) Ltd.), Mr P. Godfrey (Stanley Tools), Mr J. Davies (The Rawlplug Co. Ltd.), and especially Mr J. Ware (Kango Ltd.) and Mr P. Chamberlin (Thomas Robinson Group Plc.) for their help in checking text and artwork. I am grateful to the following organisations for supplying technical information and, where noted, for their kind permission to reproduce photographs or illustrations:

G. Cartwright Ltd.; CIBA-GEIGY (UK) Ltd.; Crosby Windows Ltd.; Denford Machine Tools Ltd. (Fig. 5.24); Dominion Machinery Co. Ltd. (Figs 5.18 and 5.20); English Abrasive and Chemicals (EAC) Ltd. (Table 2.6); European Industrial Services Ltd. (Figs 15.5(a and b) and Table 15.4); Fidor (Fibre building boards Development Organisation) Ltd.; Forestor – forest and sawmill – Equipment (Engineers) Ltd. (Fig. 1.5); Formica Ltd.; Kango Ltd. (Figs 4.1, 4.3–4.8, 4.10–4.12 and 5.22); Neill Tools Ltd.; Rabone Chesterman Ltd.; The Rawlplug Company Ltd. (Figs 15.12–15.19); Record Marples (Woodworking Tools) Ltd. (Figs 2.25–2.27, 2.29–2.34, 2.36, 2.52–2.61, 2.73–2.78, 2.97–2.100, 2.123 and 7.9); Thomas Robinson and Son (Figs 5.18, 5.5(a), 5.7(a), 5.11(a) and 5.14); Stanley Tools Ltd. (Figs 2.28, 2.35, 2.37–2.39, 2.70, 2.83, 2.84, 2.89–2.92 and Table 2.5); Stenner of Tiverton Ltd (Figs 1.6, 1.9, 1.11–1.13); The Swedish-Finnish Timber Council (Fig. 1.7); The Timber Research and Development Association (TRADA); The Timber Trades Federation; Wadkin Plc. (Figs 5.2, 5.3, 5.5(b), 5.6, 5.7(b), 5.9, 5.11(b), 5.12, 5.16 and 5.21); G.F. Wells Ltd. (Timber Drying Engineers) (Fig. 1.26).

Tables 1.2 and 1.4 are extracted from BS 4471 and Table 1.3 is extracted from BS 5450 by kind permission of the British Standards Institution. Complete copies may be obtained by post from BSI Sales, Linford Wood, Milton Keynes, Bucks. MK14 6LE.

Finally, I must thank my wife – Hilary Yvonne – for her continual help, patience and understanding during the whole of this work.

Brian Porter
Leeds 1989

1

Timber and associated materials

1.1 Growth and structure of a tree

The life of a tree begins very much like that of any other plant – the difference being that, if the seedling survives its early stage of growth to become a sapling (young tree), it may develop into one of the largest plants in the plant kingdom.

The hazards to young trees are many and varied. Animals are responsible for the destruction of many young saplings, but this is often regarded as a natural thinning-out of an otherwise overcrowded forest, thus allowing the sapling to mature and develop into a tree of natural size and shape. Where thinning has not taken place, trees grow thin and spindly – evidence of this can be seen in any overgrown woodland where trees have had to compete for the daylight necessary for their food production.

With all natural resources which are in constant demand, there comes a time when demand outweighs supply. Fortunately, although trees require 30–100 years or more to mature, it is possible to ensure a continuing supply – provided that land is made available and felling (cutting down) is strictly controlled. This has meant that varying degrees of conservation have had to be enforced throughout some of the world's largest natural forests and has led to the development of massive man-made forests (forest farming).

Components

A tree has three main parts (Fig. 1.1):

1 the root system,
2 the stem or trunk,
3 the crown.

The roots anchor the tree firmly into the ground and, via many small root hairs, absorb water and minerals, to form *sap* (see Fig. 1.2).

The stem or trunk conducts sap from the roots, stores food, and supports the crown. Timber is cut from this part of the tree.

The crown consists of branches, twigs, and foliage (leaves). Branches and twigs are the lifelines supplying the leaves with sap.

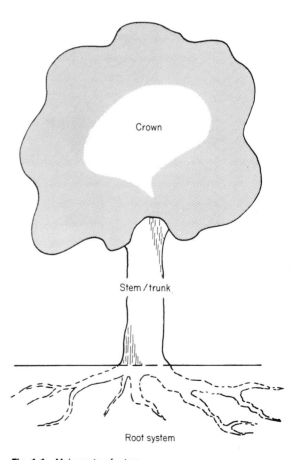

Fig. 1.1 Main parts of a tree

Food process (Fig. 1.3)

The leaves play the vital role of producing the tree's food. By absorbing daylight energy via the green pigment (chlorophyll) in the leaf, they convert a mixture of carbon dioxide taken from the air and sap from the roots into the necessary amounts of sugars and starches (referred to as *food*) while at the same time releasing oxygen into the atmosphere as a waste product. This process is known as *photosynthesis*. However, during the hours of darkness this action to some extent is reversed – the leaves take in oxygen and give off carbon dioxide, a process known as *respiration* (breathing).

For the whole process to function, there must be some form of built-in system of circulation which allows sap to rise from the ground to the leaves and then to descend as food to be distributed throughout the whole tree. It would seem that this action is due either to suction induced by *transpiration* (leaves giving off moisture by evaporation) and/or to capillarity (see Section 12.2) within the cell structure of the wood.

Structural elements

The following features are illustrated in Fig. 1.4.

Pith (medulla) – the core or centre of the tree, formed from the tree's earliest growth as a sapling.

Growth ring (sometimes referred to as an annual ring) – wood cells which have formed around the circumference of the tree during its growing season. The climate and time of year dictate the growth pattern. Each ring is often seen as two distinct bands, known as *earlywood* (springwood) and *latewood* (summerwood). Latewood is usually more dense than earlywood and can be recognised by its darker appearance.

Sapwood – the outer active part of the tree which, as its name implies, receives and conducts sap from the roots to the leaves. As this part of the tree matures, it gradually becomes heartwood.

Heartwood – the natural non-active part of the tree, often darker in colour than sapwood, gives strength and support to the tree

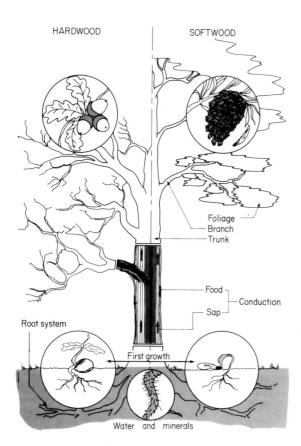

Fig. 1.2   Growth of a tree

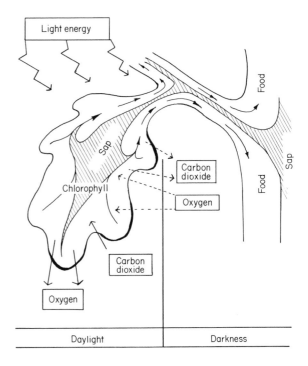

Fig. 1.3   The process of photosynthesis

and provides the most durable wood for conversion into timber.

Rays – these may all appear (although falsely – as not many do) to originate from the centre (medulla) of the tree, hence the term *medullary rays* is often used to describe this strip of cells that allow sap to percolate transversely through the wood. They are also used to store excess food.

Rays are more noticeable in hardwood than in softwood (see Section 1.2), and even then can be seen with the naked eye only in such woods as oak and beech. Fig. 1.18 shows how rays may be used as a decorative feature once the wood has been converted (sawn into timber).

Cambium – a thin layer or sleeve of cells located between the sapwood and the bast (phloem). These cells are responsible for the tree's growth. As they are formed, they become subdivided in such a way that new cells are added to both sapwood and bast, thus increasing the girth of the tree.

Inner bark, or Bast (phloem) – conducts food throughout the whole of the tree, from the leaves to the roots.

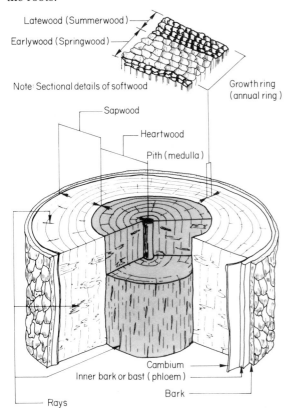

Fig. 1.4 Section through the stem/trunk

Bark – the outer sheath of the tree. It functions as

1 a moisture barrier,
2 a thermal insulator,
3 an armour plate against extremes of temperature, attack by insects and fungi, and animals.

The bark of a well established tree can usually withstand minor damage, although excessive ill treatment to this region could prove fatal.

1.2 Hardwoods and softwoods

The terms hardwood and softwood can be very confusing, as not all commercially classified hardwoods are physically hard, or softwoods soft. For example, the obeche tree is classed as a hardwood tree, yet it offers little resistance to a saw or chisel etc. The yew tree, on the other hand, is much harder to work yet it is classified as a softwood. To add to this confusion we could be led to believe that hardwood trees are deciduous (shed their leaves at the end of their growing season) and softwood trees are evergreen (retain their leaves for more than one year), which is true of most species within these groups, but not all!

Table 1.1 identifies certain characteristics found in hardwood and softwood trees; however, it should be used only as a general guide.

Table 1.1 Guide to the recognition of hardwoods and softwoods (see Fig. 1.2)

	Hardwood	Softwood (conifers)
1 Botanical group	Angiosperms	Gymnosperms
2 Leaf growth	Deciduous and evergreen	Evergreen*
3 Leaf shape	Broadleaf	Needle or scale-like leaves
4 Seed	Encased	Naked via cone
5 General use	Decorative; heavy structural	General structural

* Not always the case - e.g. larches

Hardwood and softwood in fact refer to botanical differences in cell composition and structure. (Cell types and their formation are dealt with in Carpentry and Joinery *2*.)

1.3 Sources and supply of timber

The forests of the world that supply the wood for

timber, veneers, wood pulp, and chippings for particle board are usually situated in areas which are typical for a particular group of tree species. For example, the coniferous forests supplying the bulk of the world's softwoods are mainly found in the cooler regions of northern Europe, also Canada and Asia – stretching to the edge of the Arctic Circle. Hardwoods, however, come either from a temperate climate (neither very hot nor very cold) – where they are mixed with faster-growing softwoods – or from subtropical and tropical regions, where a variety of hardwoods grow. Their type being specific to different continents.

Tree and timber names

Common names are often given to trees (and other plants) so as to include a group of similar yet botanically different species. It is these common English names which are predominantly used in our timber industry. The true name or Latin botanical name of the tree must be used where formal identification is required – for example:

Common English name	Species (true name or Latin botanical name)	
	Genus (generic name or 'surname')	Specific name (or 'forename')
Scots pine	*Pinus*	*sylvestris*

As a general guide, it could therefore be said that plants have both a surname and a forename and, to take it a step further, belong to family groups of hardwood and softwood.

Commercial names for timber often cover more than one species. In these cases, the botanical grouping is indicated by *spp.*.

Softwoods

Most timber used in the UK for carpentry and joinery purposes is softwood imported from Sweden, Finland, and the USSR. The most important of these softwoods are European redwood (*Pinus sylvestris*), which includes Baltic redwood, and Scots pine, a native of the British Isles. As timber, these softwoods are collectively called simply 'redwood'. Redwood is closely followed in popularity by European whitewood, a group which includes Baltic whitewood and Norway spruce (*Picia abies*) – recognised the world over as the tree most commonly used at Christmas

time. Commercially these and sometimes silver firs are simply referred to as 'whitewood'.

Larger growing softwoods are found in the Pacific coast region of the USA and Canada. These include such species as Douglas fir (*Pseudotsuga menziesii*) – known also as Columbian or Oregon pine, although technically not a pine – Western hemlock (*Tsuga heterophylla*), and Western red cedar (*Thuja plicata*) which is known for its durability, being almost immune from attack by insects or fungi. Brazil is the home of Parana pine (*Araucaria angustifolia*) which produces long lengths of virtually knot-free timber which is, however, only suitable for interior joinery purposes.

Temperate hardwoods

These hardwood trees are found where the climate is of a temperate nature. The temperate regions stretch north and south from the tropical areas of the world, into the USSR, Europe, China and North America in the northern hemisphere and Australia, New Zealand and South America in the southern hemisphere.

The United Kingdom is host to many of these trees, but not in sufficient quantities to meet all its needs. It must therefore rely on imports from countries which can provide such species as oak (*Quercus* spp.), sycamore (*Acer* spp.), ash (*Fraxinus* spp.), birch (*Betula* spp.), beech (*Fagus* spp.), and elm (*Ulmus* spp.) – which is now an endangered species due to Dutch elm disease.

Tropical and subtropical hardwoods

Most tropical hardwoods come from the rain forests of South America, Africa, and South East Asia. Listed below are some hardwoods which are commonly used:

African mahogany (*Khaya* spp.)	– West Africa
Afrormosia (*Pericopsis elata*)	– West Africa
Agba (*Gossweilerodendron balsamiferum*)	– West Africa
American mahogany (Brazilian) (*Swietenia macrophylla*)	– Central and South America
Gaboon (*Aucoumea klaineana*)	– West Africa
Iroko (*Chlorophora excelsa*)	– West Africa

Keruing (*Dipterocarpus*   – South East Asia
spp.)
Meranti (*Shorea* spp.)   – South East Asia
Ramin   – South East Asia
(*Gonostylus* spp.)
Sapele   – West Africa
(*Entandrophragma*
cylindricum)
Teak   – Burma, Thailand
(*Tectona grandis*)
Utile (*Entandophragma*   – West Africa
utile)

Hardwood use

Hardwoods may be placed in one or more of the
following groups:

Purpose group	*Use*
a) Decorative	Natural beauty – colour and/or figured grain
b) General-purpose	Joinery and light structural
c) Heavy structural	Withstanding heavy loads

Forms of supply

Softwood is usually exported from its country of
origin as sawn timber in packages or in bundles. It
has usually been pre-dried to about 20% m.c.
(moisture content). Packaged timber is to a
specified quality and size, bound or bonded with
straps of steel or plastics for easy handling, and
wrapped in paper or plastics sheets.

Hardwood, however, may be supplied as sawn
boards or as logs to be converted (sawn) later by
the timber importer to suit the customer's
requirements.

1.4 Conversion

Felling (the act of cutting down a living tree) is
carried out when trees are of a commercially
suitable size, having reached maturity, or for
thinning-out purposes. Once the tree has been
felled, its branches will be removed, leaving the
trunk in the form of a log. This log is the portion of
the tree which is broken down by being sawn
(converted) into timber – hence the term *conversion*.

What, then, is the difference between *wood* and
timber? The word wood is often used very loosely to
describe timber, when it should be used to describe
either a collection of growing trees or the substance
that trees are made of, i.e. the moisture-conducting

cells and tissues etc. Timber is wood in the form of
squared boards or planks etc.

Initial conversion may be carried out in the
forest, using heavy yet portable machines such as
circular saws or vertical and horizontal band-mills
(see Fig. 1.5). This leads to a reduction in
transport cost, as squared sectional timber can be
transported more economically than logs.

Alternatively, the logs may be transported by
road, rail, or water to a permanently sited sawmill.

The type of sawing equipment used in a sawmill
will depend on the size and kind of logs it handles.
For example:

a) *Circular saw* (Fig. 1.6) – small- to medium-
diameter hardwoods and softwoods. Figure 1.6
shows a rolling table log saw. The tables are
available in lengths from 3.05 m to 15.24 m,
and the diameter of saw could be as large as
1.829 m.

b) *Vertical frame saw or gang saw* (Fig. 1.7) – small-
to medium-diameter softwoods. The log is fed
and held in position by fluted rollers while
being cut with a series of reciprocating
(upward-and-downward moving) saw blades.
The number and position of these blades will
vary according to the size and shape of each
timber section. Fig. 1.8(a) illustrates the
possible result after having passed the log
through this machine once, whereas
Fig. 1.8(b) shows what could be achieved after
making a further pass.

c) *Vertical band-mill* (Fig. 1.9) – all sizes of both
hardwood and softwood. Logs are fed by a
mechanised carriage to a saw blade in the form
of an endless band which revolves around two

Fig. 1.5 'Forestor-150' horizontal band-mill - through-and-
through sawing

Fig. 1.6   Circular saw

Fig. 1.7   Frame saw

Fig. 1.9   Band-mill capable of cutting logs up to 1.220 m diameter

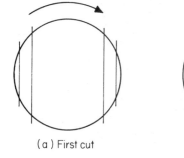

(a) First cut   (b) Second cut

Fig. 1.8   Frame saw cut

large wheels (pulleys), one of which is motorised. Figure 1.10 shows an example of how these cuts can be taken.

d) *Double vertical band-saw* (Fig. 1.11) – small to medium logs. It has the advantage of making two cuts in one pass.

e) *Horizontal band-saw* (Fig. 1.5) – all sizes of hardwood and softwood. The machine illustrated is suitable for work at the forest site, or in a sawmill. Conversion is achieved by passing the whole mobile saw unit (which travels on rails) over a stationary log, taking a slice off at each forward pass.

The larger mills may employ a semi-computerised system of controls to their machinery, thus helping to cut down some human error and providing greater safety to the whole operation. The final control and decisions, however, are usually left to the expertise of the sawyer (machine operator).

Timber which requires further reduction in size is cut on a resaw machine. Figure 1.12 shows a resaw operation being carried out, and Fig. 1.13 illustrates how two machines can be employed to speed up the operation.

Importers of timber in the United Kingdom may specialise in either hardwoods or softwoods or both. Their sawmills will be geared to meet their particular needs, by resawing to customers' requirements. Hardwood specialists usually have their own drying facilities.

Timber sections

The way in which the log is cut (subdivided) will depend on the following factors:

a) cross-sectional area,
b) type of wood,
c) condition of the wood – structural defects etc. (see Section 1.6),
d) proportion of heartwood to sapwood,
e) future use – structural, decorative, or both.

Broadly speaking, the measures taken to meet the customer's requirements will be the responsibility of the experienced sawyer (as mentioned earlier). His decisions will determine the method of conversion.

Through-and-through-sawn (Fig. 1.14) In this method of conversion, parallel cuts are made down the length of the log, producing a number of 'radial' and 'tangential' sawn boards (Figs 1.15 and 1.16). The first and last cuts leave a portion of wood called a 'slab'. This method of conversion is probably the simplest and least expensive.

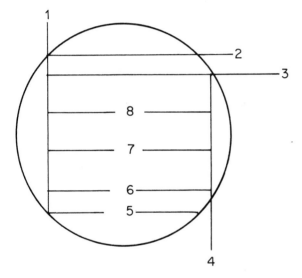

Fig. 1.10 Band-mill cuts

Fig. 1.11 Double vertical band saw

Bandsaw blades

Tangential-sawn (Plain sawn)
(Fig. 1.15)   Starting with a squared log, tangential-sawn boards are produced by working round the log by turning it, to produce boards all of which (except the centre) have their growth rings across the boards' width. Although tangential-sawn sections are subject to cupping (becoming hollow across the width) when they dry, they are the most suitable sections for softwood beams, i.e. floor joists, roof rafters, etc., which rely on the position of the growth ring to give greater strength to the beam's depth.

Quarter (radial) or rift-sawn (Fig. 1.16)   This method of conversion can be wasteful and expensive, although it is necessary where a large number of radial or near radial-sawn boards are required. Certain hardwoods cut in this fashion

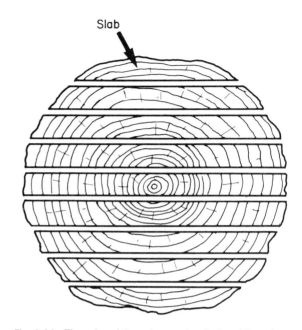

Fig. 1.14   Through-and-through sawn (producing plain and quarter sawn timber)

Fig. 1.12   Resaw operation

Fig. 1.13   Band resaw in tandem operation

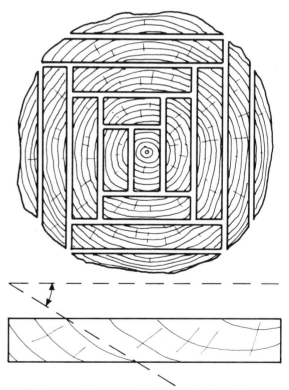

Plain sawn timber – growth rings meet the face of the board at an angle less than 45%

Fig. 1.15   Tangential sawn (producing plain sawn timber - except for heartboards)

produce beautiful figured boards (Fig. 1.18) for example figured oak, as a result of the rays being exposed (Fig. 1.4). Quarter-sawn boards retain their shape better than tangential-sawn boards

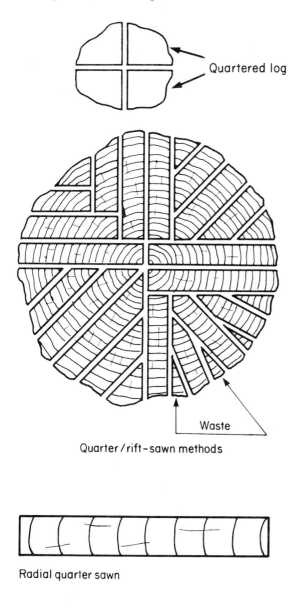

Quarter/rift-sawn methods

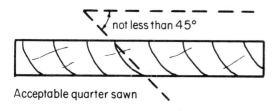

Radial quarter sawn

Acceptable quarter sawn

Fig. 1.16 Quarter sawn

and tend to shrink less, making them well suited to good-class joinery work and quality flooring.

Conversion geometry (Fig. 1.17) Knowing that a log's cross-section is generally just about circular, the above-mentioned saw cuts and sections can be related to a circle and its geometry. For example, timber sawn from a 'radius' line will be *radial-sawn* or quartered logs (divided by cutting into four quarters, or *quadrants*). Similarly, any cut made as a tangent to a growth ring would be called *tangential-sawn*. The chords are straight lines which start and finish at the circumference; therefore a series of chords can be related to a log which has been sawn through-and-through or plain sawn. It should be noted that the chord line is also used when cuts are made tangential to a growth ring, and when the log is cut in half.

Decorative-sawn boards Fig. 1.18 gives two examples of how wood can be cut to produce timber with an attractive face.

Plain sawn softwood relies on the appearance of its growth rings, whereas the hardwood is quarter-sawn to show off its rays to great advantage.

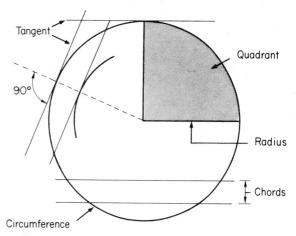

Fig. 1.17 Conversion geometry

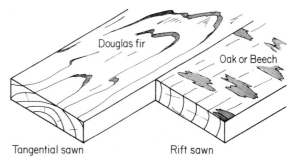

Fig. 1.18 Grain figuring

1.5 Size and selection of timber

Sawn timber is available in a wide variety of lengths, cross-sections, and species, to meet the different needs of the construction and building industry. Tables 1.2 and 1.3 show some of the standard sizes that can be made available through timber merchants. Figure 1.19 shows the sequence and method of ordering. The species, actual cross-section, length, quantity, and processing (or finish) should be quoted in that order.

Table 1.2 Basic sizes of sawn softwood

a) Cross-sectional sizes (mm)

Thickness	Width								
	75	100	125	150	175	200	225	250	300
16	×	×	×	×					
19	×	×	×	×					
22	×	×	×	×	×	×	×		
25	×	×	×	×	×	×	×	×	×
32	×	×	×	×	×	×	×	×	×
36	×	×	×	×					
38	×	×	×	×	×	×	×	×	×
44	×	×	×	×	×	×	×	×	×
47	×	×	×	×	×	×	×	×	×
50	×	×	×	×	×	×	×	×	×
63		×	×	×	×	×	×		
75		×	×	×	×	×	×	×	×
100		×		×		×	×	×	×
150				×		×			×
200						×			
250								×	
300									×

Note Certain sizes may not be obtainable in the normal range of species and grades which are generally available.

b) Lengths (m)

1.80	2.10	3.00	4.20	5.10	6.00	7.20
	2.40	3.30	4.50	5.40	6.30	
	2.70	3.60	4.80	5.70	6.60	
		3.90			6.90	

Note Lengths of 6.00 m and over may not be readily available.

Table 1.3 Basic sizes of sawn hardwood

a) Cross-sectional sizes (mm)

Thickness	Width										
	50	63	75	100	125	150	175	200	225	250	300
mm											
19			×	×	×	×	×				
25	×	×	×	×	×	×	×	×	×	×	×
32			×	×	×	×	×	×	×	×	×
38			×	×	×	×	×	×	×	×	×
50				×	×	×	×	×	×	×	×
63						×	×	×	×	×	×
75					×	×	×	×	×	×	×
100					×	×	×	×	×	×	×

Note Designers and users should check the availability of specified sizes in any particular species.

b) Lengths The basic lengths of hardwoods shall be any integral multiple of 100 mm, but not less than 1 m.
Note The normal length of imported hardwood will vary according to the species and the origin.

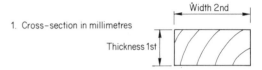

1. Cross-section in millimetres

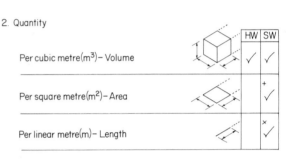

2. Quantity

	HW	SW
Per cubic metre(m³) – Volume	✓	✓
Per square metre(m²) – Area	+	✓
Per linear metre(m) – Length	×	✓

+ e.g. Floor or matchboard etc.
× e.g. Small quantities

3. Finish
   Type of processing (Fig 1.20)

Fig. 1.19 Sequence and method of planing

By adopting standard sizes, we can reduce the time spent on further conversion, subsequent wastage, and the inevitable build-up of short ends or *off-cuts* (off-cuts usually refers to waste pieces of sheet materials), thereby making it possible to plan jobs more efficiently and economically.

A large part of our industry is made up of small joinery firms which, due to their size, do not always have the machinery or storage facilities to handle large quantities of timber. These firms rely on small timber merchants to provide a service whereby stock sizes of both sawn and planed timber (Table 1.4) are readily available.

Buying pre-planed or shaped timber adds considerably to the cost – further stressing the importance of selecting the correct standard stock sizes wherever possible.

Figure 1.20 shows how the sizes in Table 1.4 have been achieved, together with the planing sequence as follows:

a) EX   –   sawn to nominal size

b) S1S – surfaced one
side or P1S – planed one side

c) S1S1E – surfaced one side and one
edge or P1S1E – planed one
side and one edge

d) S1S2E – surfaced one side and two
edges or P1S2E – planed one
side and two edges

e) S4S – surfaced four
sides or P4S – planed four
sides

p.a.r. – planed all round
or p.s.e. – planed square edged

Being able to recognise a piece of timber by its common name can be just as important as ordering the correct amounts. Listed below is a selection of common-named timbers. By referring to Sections 1.2 and 1.3, see if you can allocate them to their botanical classification, i.e. HW (hardwood) or SW (softwood).

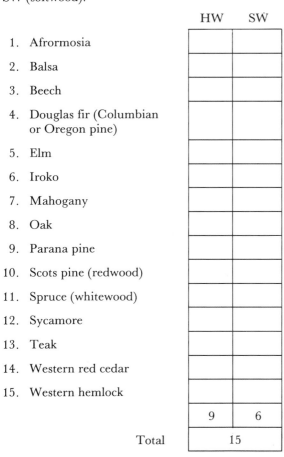

	HW	SW
1. Afrormosia		
2. Balsa		
3. Beech		
4. Douglas fir (Columbian or Oregon pine)		
5. Elm		
6. Iroko		
7. Mahogany		
8. Oak		
9. Parana pine		
10. Scots pine (redwood)		
11. Spruce (whitewood)		
12. Sycamore		
13. Teak		
14. Western red cedar		
15. Western hemlock		
	9	6
Total		15

[*Answers:* HW – 1, 2, 3, 5, 6, 7, 8, 12, 13; SW – 4, 9, 10, 11, 14, 15]

Common timbers are usually identified by recognising familiar characteristics, for example colour, weight, texture, and smell.

Table 1.4 Reductions of sawn softwood from basic nominal size to finished size by processing of two opposed faces

	Reduction from basic size to finished size				
	For basic sawn sizes of width or thickness (mm):				
End use or product	15 to 25	26 to 50	51 to 100	101 to 150	151 to 300
Constructional timber surfaced	3	3	3	5	6
Floorings, matchings and interlocked boarding and planed all round	5	6	7	7	7
	5	6	7	7	7
Trim	6	7	8	9	10
Joinery and cabinetwork	7	9	10	12	14

1.6 Structural defects (natural defects)

Figures 1.21 to 1.23 show defects that may be evident before and/or during conversion. Most of these defects have little, if any, detrimental effect on the tree, but they can degrade the timber cut from it, i.e. lower its market value.

Heart shake (Fig. 1.21(a)) – shake (parting of wood fibres along the grain) within the heart (area around the pith) of the tree caused by uneven stresses, which increase as the wood dries.

Star shake (Fig. 1.21(a)) – a collection of shakes radiating from the heart.

Ring shake (cup shake) (Fig. 1.21(b)) – a shake which follows the path of a growth ring. Figure 1.21(c) shows the effect it can have on a length of timber.

Rate of growth (Fig. 1.21(d)) – the number of growth rings per 25 mm determines the strength of the timber.

Compression failure (upset) (Fig. 1.21(e)) – fracturing of the fibres; thought to be caused by sudden shock at the time of felling or by the tree becoming overstressed (during growth) – possibly due to strong winds etc. Found mainly in mahogany, and meranti (shorea spp).

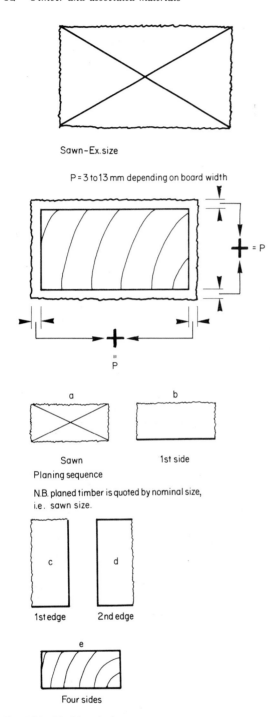

Sawn-Ex.size

P = 3 to 13 mm depending on board width

= P

"
P

| a | b |

Sawn   1st side

Planing sequence

N.B. planed timber is quoted by nominal size, i.e. sawn size.

| c | d |

1st edge   2nd edge

e

Four sides

Fig. 1.20 Machine planing

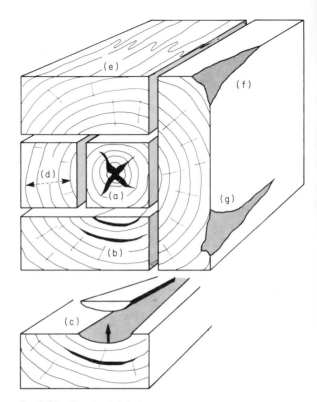

Fig. 1.21 Structural defects

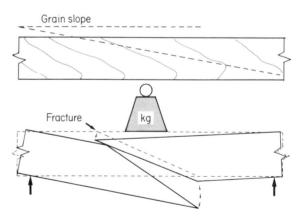

Grain slope

Fracture

kg

Fig. 1.22 Sloping grain - possible faults

Wane (waney edge) (Fig. 1.21(f)) – the edge of a piece of timber which has retained part of the tree's rounded surface.

Encased bark (Fig. 1.21(g)) – may appear on the face or the edge of a piece of timber.

Sloping grain (Fig. 1.22) – the grain (direction of the wood fibres) slopes in a way that can make load-bearing timbers unsafe, e.g. beams and joists.

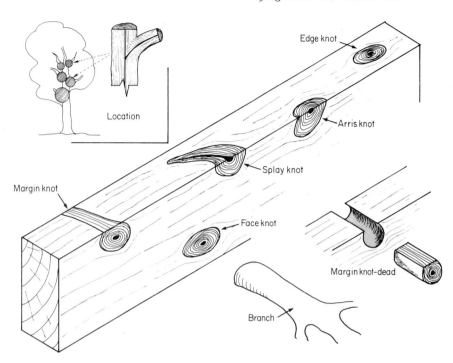

Fig. 1.23 Knot recognition - systems of knots

Knots (Fig. 1.23) – where the tree's branches have joined the stem and become an integral part of it. Figure 1.23 shows how knots may appear in the sawn timber. The size, type, location, and number of knots are controlling factors when the timber is graded for use.

Dead knots (Fig. 1.23)   If a branch is severely damaged, that part adjoining the stem will die and may eventually become enclosed as the tree

develops – not being revealed until conversion into timber. Note: these knots are often loose, making them a potential hazard whenever machining operations are carried out.

1.7 Drying timber and moisture content

Timber derived from freshly felled wood is said to be *green*, meaning that the cell cavities contain *free water* and the wall fibres are saturated with bound water (Fig. 1.28), making the wood heavy, structurally weak, susceptible to attack by insects and/or fungi, and unworkable. Timber in this condition is therefore always unsuitable for use. The amount of moisture the wood contains as a percentage of the oven-dry weight is known as the *moisture content* (m.c.), and the process of reducing the m.c. is termed *drying*.

Drying (sometimes called *seasoning*), timber is usually carried out by one of three methods:

1   air drying (natural drying),
2   kiln drying (artificial drying),
3   air drying followed by kiln drying.

All three methods aim at producing timber that will remain stable in both size and shape – the overriding factor being the final moisture content, which ultimately controls the use of the timber. Some examples can be seen in Fig. 1.24.

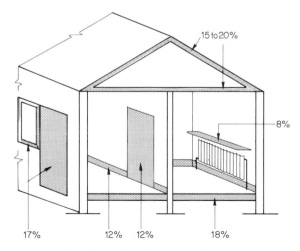

N.B. Wood with a 20% + M.C. is liable to attack by fungi

Fig. 1.24 Moisture content of timber in various situations

Air drying

Air drying is carried out in open-sided sheds, where the timber is exposed to the combined action of circulating air and temperature, which lifts and drives away unwanted moisture by a process of evaporation (similar to the drying of clothes on a washing line). A suitable reduction in m.c. can take many months, depending on

a)   the amount of exposure,
b)   the type of wood (hardwood or softwood),
c)   the particular species,
d)   the timber thickness.

The final m.c. obtained can be as low as 16% to 17% in summer months and as high as 20% or more during winter. It would therefore be fair to say that this method of drying timber is very unreliable.

A typical arrangement for air drying is shown in Fig. 1.25, where the features numbered are of prime importance if satisfactory results are to be achieved. They are as follows:

1   Timber stacks (piles of sawn timber) must always be raised off the floor, thus avoiding rising damp from the ground. Concrete, gravel, or ash will provide a suitable site covering.

2   The area surrounding the shed must be kept free from ground vegetation, to avoid conduction of moisture from the ground.
3   Free circulation of air must be maintained throughout the stack – the size and position of 'sticks' will depend on the type, species, and section of timber being dried.
4   The roof covering must be sound, to protect the stacks from adverse weather conditions.

Kiln drying

This method of drying timber speeds up the drying process from months to days and produces an accurate and uniform m.c. throughout the whole stack.

Drying kilns may vary in their construction and methods of raising heat, but the working principles remain the same – to provide fully controllable drying conditions that will meet the requirements of different timbers without 'degrading' them in any way. (Defects caused by drying are discussed in Carpentry and Joinery 2).

These drying kilns must therefore be provided with means of producing

a)   enough heat   –   to sustain the required drying temperature,
b)   enough moisture   –   so that the correct level of humidity can be maintained,
c)   enough air circulation   –   to carry away moisture and ensure even drying throughout the complete stack.

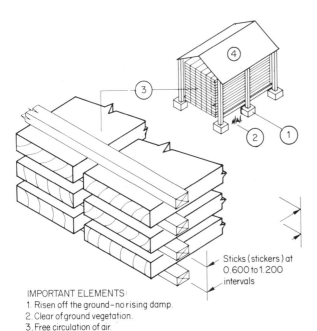

IMPORTANT ELEMENTS:
1. Risen off the ground – no rising damp.
2. Clear of ground vegetation.
3. Free circulation of air.
4. Protection from the weather.

Sticks (stickers) at 0.600 to 1.200 intervals

Fig. 1.25   Air drying

Fig. 1.26   Typical loading of timber into a kiln

The medium most commonly used to satisfy requirements (a) and (b) is *steam*.

Figure 1.26 shows a fully sticked stack of timber on its trolley, ready to be loaded into a kiln. The emission of vapour from the vents towards the top right-hand side of the picture indicates that the first two chambers are in operation. The cut-away section in Fig. 1.27 shows a typical arrangement for circulating the air around the chamber, thus encouraging the evaporation of unwanted moisture from the wood and the subsequent drying process.

Moisture content

Figure 1.28 illustrates how moisture is lost and the effect on a timber section when it is present. If a timber section is to retain its desired m.c. after having gone through all the necessary drying stages, it must be kept in an environment conducive to its m.c. level, because wood is hygroscopic (having the ability to absorb moisture from the atmosphere). Any increase in moisture will result in the wood swelling.

It is, however, possible for timber in a changeable environment to remain stable if moisture absorption can be prevented. This can be

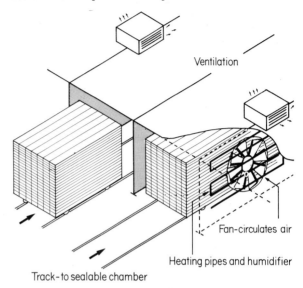

Ventilation

Fan-circulates air

Heating pipes and humidifier

Track-to sealable chamber

Fig. 1.27 Kiln drying (doors not shown)

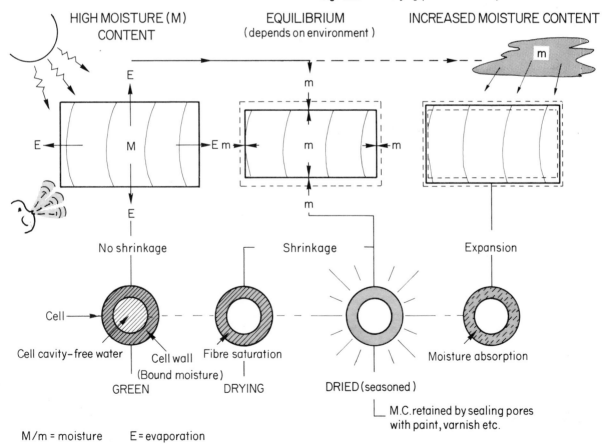

HIGH MOISTURE (M) CONTENT

EQUILIBRIUM (depends on environment)

INCREASED MOISTURE CONTENT

No shrinkage

Shrinkage

Expansion

Cell

Cell cavity–free water

Cell wall (Bound moisture)

Fibre saturation

Moisture absorption

GREEN

DRYING

DRIED (seasoned)

M.C. retained by sealing pores with paint, varnish etc.

M/m = moisture   E = evaporation

Fig. 1.28 Moisture movement

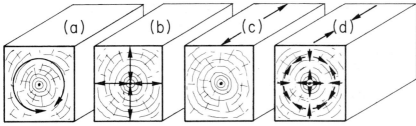

(a) Tangentially-greatest amount
(b) Radially-about half of 'a'
(c) Length-least amount
(d) Overall shrinkage

Fig. 1.29   Proportion of wood shrinkage

achieved by one of two methods:

1   completely sealing all its exposed surfaces,
2   using a micro-pore sealer which prevents direct entry of water from outside but allows trapped moisture to escape.

All timber must of course be suitably dried before the above treatments are carried out.

Shrinkage

The loss of moisture from the wood during the drying process should now be understood. However, this loss of moisture can have quite an adverse effect on the size and shape of the timber.

The proportion of movement (shrinkage) that takes place is shown in Fig. 1.29, where it will be seen that the greatest amount of movement takes place tangentially, that is to say in the direction of each growth ring. Radial shrinkage accounts for about half that amount, whereas shrinkage along the length of the grain will be least of all. By relating Fig. 1.29(a) to Fig. 1.29(d), it should be possible to see why and how timber sections become distorted when dried.

Fig. 1.30 gives examples of how different cuts from the log become affected by shrinkage.

1.8 Common enemies

If its condition is favourable, most wood can be attacked by fungi or wood-boring beetles or both. Such attacks are often responsible for the destruction of many of our trees (e.g. Dutch elm disease) and for the decomposition and subsequent failure of many timbers commonly used in building.

Fungi

There are many different types and species of fungi, all belonging to the plant kingdom. The fungi that concern us are those that take their food from and live on or in growing trees and the timber cut from them.

'Sap-staining' fungi, often known as 'blue-stain' or 'blueing' fungi, obtain their nourishment from the cell contents of sapwood, leaving the cell walls unscathed; therefore the main degrading effect these fungi have on wood is the bluish discolouration they leave. They only attack wood with a high m.c.; so, to prevent attacks by these fungi, it is necessary that conversion and drying are carried out as early as possible after the trees have been felled.

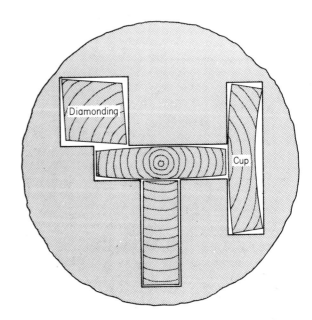

Fig. 1.30   Shrinkage - its possible effect on timber

However, the wood-destroying types of fungi are a different matter, as they live off the cell walls of the wood, thus causing its whole structure to decompose and eventually collapse. Their growth requirements are simply wood with above 20% m.c. for their initial germination and subsequent nourishment in the form of non-durable sapwood.

Durable timbers are those which have a natural resistance to fungal attack; however, the majority of timbers have either to be kept permanently below 20% m.c. or to undergo treatment with a suitable wood preservative, as an assurance against attack.

All wood-destroying fungi have a similar life cycle, and a typical example is shown in Fig. 1.31. The spores have been transported from the parent plant (fruiting body or *sporophore*) – by wind, insects, animals, or an unsuspecting human – to a suitable piece of fertile wood where germination can take place. Once established, the fungus spreads its roots (hyphae) – which are in fact the body of the fungus – into and along the wood in search of food, eventually to become a mass of tubular threads which collectively are called *mycelium*, which will produce a fruiting body.

Fruiting bodies (sporophores) take the shape of stalks, brackets, or plates etc. (Fig. 1.32), depending on the species of fungus. Each fruiting body is capable of producing and shedding millions of minute spores, of which only a very small proportion will germinate.

Wood-boring beetles

The term *wood-boring beetle* as opposed to woodworm can be confusing. Although this beetle is capable of biting holes into wood, it is the larva or grub (woodworm) of the beetle which is directly responsible for the damage done to the wood. Damage is brought about by the endless tunnelling of the larva while it feeds on the wood substance. It is therefore inevitable that wherever mass infestation is present there will always be the danger that the whole of the wood structure may collapse.

There are many varieties of wood-boring beetle – each with its own life style. For example, there are those which attack living trees, or which prefer trees which have been recently felled, while

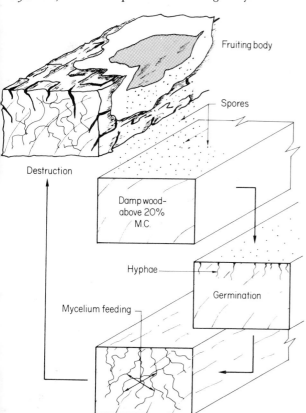

Fig. 1.31 Process of wood-destroying fungi

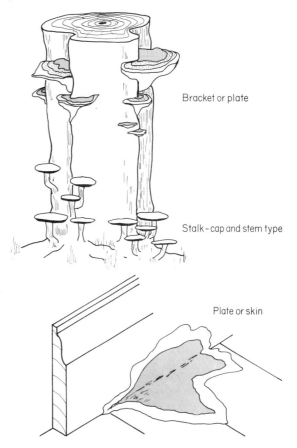

Fig. 1.32 Fruiting bodies (sporophores)

others only attack certain species – and there are those which are not too particular.

Probably the best known beetle in the United Kingdom is the *common furniture beetle*. Evidence of its attack can be found in many homes, in both furniture and structural members, where it devours hardwood, softwood, and plywood alike, although it does tend to prefer the sapwood first.

Figure 1.33 represents a typical life cycle of a wood-destroying beetle:

1   Eggs will be laid by the female beetle in small cavities below the surface of the wood.
2   After a short period (usually a few weeks) the eggs hatch into larvae (grubs) and enter the wood, where they progressively gnaw their way further into the wood, leaving excreted wood dust (frass) in the tunnel as they move along.
3   After one or more years of tunnelling (depending on the species of beetle) the larva undermines a small chamber just below the surface of the wood, where it *pupates* (turns into a chrysalis).
4   The *pupa* then takes the form of a beetle and emerges from its chamber by biting its way out, leaving a hole known as an *exit* or *flight* hole. (Collectively, these holes are usually the first sign of any insect attack.) The insect is now free to travel or fly at will – to mate and complete its life cycle.

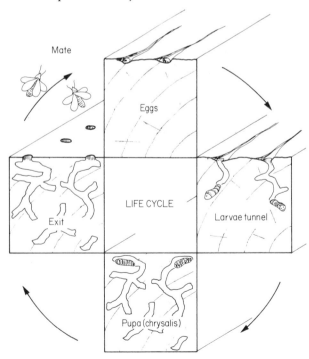

Fig. 1.33 Life cycle of a wood-boring beetle

Wood-destroying beetles can often be identified by one or more of the following characteristics;

a)   habitat,
b)   size and shape of the beetle,
c)   size and shape of the larva,
d)   size and shape of the frass,
e)   size and shape of the flight holes,
f)   sound.

Probably the best preventative measure against insect attack is, where practicable, to keep the m.c. of wood below 10%, for at this level these insects will be discouraged from breeding. As can be seen from Fig. 1.24, most structural timber members of a building exceed this m.c., therefore if the building has a history of insect attack or is sited in a geographical area where infestation is common (as is the case in certain areas in the United Kingdom) then suitable preservative methods should be considered.

Preventative and remedial treatment for outbreaks of both fungal and insect attack are dealt with in *Carpentry and Joinery 2*.

1.9   Manufactured boards

We have seen how wood can be subject to dimensional change and distortion when used in its solid state. It is this inherent problem, together with its cost, that often restricts the use of wood where wide or large areas have to be covered. This is the kind of work where manufactured boards are mainly used.

For the purpose of this chapter, 'manufactured boards' will be taken to mean those sheet materials which for their greater part are composed of wood veneer, strips, particles, or their combination. They fall into the following groups:

1   Plywoods:
    a)   veneer plywood,
    b)   core plywood-laminated boads (laminboard, blockboard, battenboard).
2   Particle boards:
    a)   chipboards,
    b)   wood cement particle board.
3   Flake boards:
    a)   waferboard
    b)   OSB (orientated strand board).
4   Fibre boards:
    a)   insulation board,
    b)   medium board,
    c)   medium density fibreboard (MDF),
    d)   hardboard.

Veneer plywood (Fig. 1.34)

The word *plywood* is usually taken to refer to those sheets or boards which are made from three or more odd numbers of thin layers of wood – known as wood *veneers*. It is important that all veneers on each side of the core or centre veneer are balanced (see Fig. 1.42).

Three ply (Fig. 1.34(a))   consists of a face and back veneer (ply) sandwiching either a central veneer of the same thickness or a core of thicker veneer, usually of a lesser density (Fig. 1.34(b)).

Multi-ply (Fig. 1.34(c)) – the face, core, and backing consist of more than three plies, usually of a similar thickness.

Plywood veneers are usually produced by peeling a log (Fig. 1.35). The logs are cut to length, debarked, and (depending on the species) often steamed or given a similar treatment to soften the wood fibres before being peeled along their length. The resulting veneers are then cut to length, dried, and (depending on the thickness and number of plys required) glued together with the grain directions of alternate veneers running at right angles to one another. They are then pressed, cured, dried, trimmed to size, and finally dressed by sanding.

Note: outer decorative veneers are usually cut by

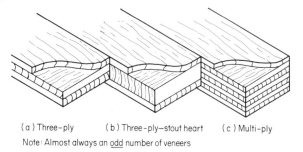

(a) Three-ply   (b) Three-ply–stout heart   (c) Multi-ply
Note: Almost always an <u>odd</u> number of veneers

Fig. 1.34   Plywood

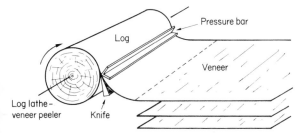

Fig. 1.35   Rotary veneer cutting

slicing them from the face of a *flitch* (a large piece of square or bevel-edged timber cut from the log).

The type of glue used to bond the veneers together classifies the plywood with regard to its type and general use, as shown in Table 1.5.

Table 1.5   Plywood adhesives and use

Adhesive type (bonding agent)	Plywood type	Use
INT	Interior use only	Not to be subjected to dampness
MR	Moisture-resistant	General use
CBR	Cyclical boil-resistant	Will not withstand extreme weather conditions
WBP	Weather-and boil-proof	Exterior quality

Although the adhesives used to bond the veneers control the plywood's use, the quality and final grade of the plywood is usually determined by the condition of the outer veneers, i.e. the presence of blemishes and knot holes etc., and the durability rating of the veneers.

Uses of veneer plywood   Depending on type and quality, plywood is used in a variety of ways, for example as

a)   formwork,
b)   roof decking,
c)   flooring,
d)   sheathing (covering the walls of timber-framed houses),
e)   cladding,
f)   wall panelling,
g)   ceilings,
h)   door panels,
i)   carcassing (or frameworks),
j)   shelving.

Core plywoods (laminated boards) (Figs 1.36 – 1.38)

Laminboard (Fig. 1.36)   This has a core made up of a lamination of narrow wood strips – not exceeding 7 mm wide – glued together, then faced with one or two veneers on each face. Laminboard has a virtually distortion-free surface.

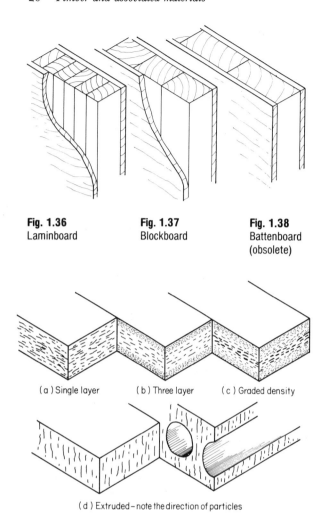

Fig. 1.36
Laminboard

Fig. 1.37
Blockboard

Fig. 1.38
Battenboard
(obsolete)

(a) Single layer   (b) Three layer   (c) Graded density

(d) Extruded – note the direction of particles

Fig. 1.39   Particle board

Blockboard (Fig. 1.37)   This is similar to
laminboard, except that the wood strips used in its
core are wider – usually between 19 mm and
28 mm. The board's finished surface is often
slightly rippled.

Battenboard (Fig. 1.38)   The only significant
difference here is that the core material is much
wider. It therefore follows that this type of board
will be less stable than the others, and subject to
more irregularities.

Battenboard is now obsolete and therefore not
often seen in the United Kingdom.

Uses of laminated boards   The use of laminated
boards is restricted to those areas which are not
subjected to dampness, as most of these boards are
bonded together with INT types of adhesives.

Particle board (Fig. 1.39)

The main natural ingredients which go to make
particle board are wood chippings; hence the
common name of *chipboard*.

Wood required to produce these chippings
comes from many different sources:

a)   young trees,
b)   forest thinnings,
c)   slabs from sawmills (Fig. 1.14),
d)   wood-machining waste – shavings, chippings,
etc.

Fibres from the flax plant also provide valuable
raw material for particle board production.

In simple terms, the manufacturing process
involves the shredding of the raw material; drying
and mixing it with a suitable synthetic-resin
adhesive; and then, with the exception of *extruded*
boards, the substance is pressed flat between
platens of a hot press, trimmed to size, and allowed
to mature; then finally finished by sanding.

There are several types of these pressed boards,
such as

a)   single-layer,
b)   three-layer,
c)   multi-layer,
d)   graded-density.

Single-layer boards (Fig. 1.39(a)) – a uniform
mass of particles of either wood, flax, or both. The
type, grade, and compaction of these particles will
affect the board's strength and working
properties – boards may for instance be classed as
interior structural with regard to their use.

Because of their composition (uniformity of
particles), single-layer boards present very few
problems when being cut.

Three-layer boards (Fig. 1.39(b)) – these consist
of a low-density core of large particles, sandwiched
between two relatively higher-density layers of fine
particles. These boards have a very smooth even
surface suitable for direct painting etc. They may
be classed as general-purpose or as interior non-
structural.

Multi-layer boards (not illustrated) – similar to
three-layer but for an increase in the number of
layers, and possibly a layer of higher density core
to improve strength.

Care should be taken when cutting these boards,
as, unlike single-layer types, they are inclined to
split or chip away at the cut edge.

Graded-density boards (Fig. 1.39(c)) – these have a board structure midway between the single-layer and three-layer types. Their particles vary in size, getting smaller from the centre outwards. They are suitable for non-structural use and for furniture production.

Extruded boards (Fig. 1.39(d)) – the prepared mixture of shredded chippings and adhesive passes through a die, resulting in an extruded board of predetermined thickness and width but of unlimited length. The holes in these boards are made by metal heating tubes, which assist in curing the adhesive, thus enabling much thicker boards to be produced. These holes also reduce the overall weight of the board.

Chipboard produced in this way will have some of its particles located at right angles to the face of the board, thus reducing its strength. However, the main use of these boards is as comparatively lightweight core material to be sandwiched between suitable layers of veneers or other materials to give it the required stability.

Uses of chipboard   Chipboard generally can be used in situations similar to plywood. Its use is however determined by

1   its method of manufacture,
2   the bonding agent,
3   special treatment – surface (e.g. laminates) or integral (e.g. fire retardants).

BS 5669 defines four grades of chipboard for particular uses:

Standard (*Type I*)	– general-purpose, e.g. furniture production.
Flooring (*Type II*)	– general domestic flooring grade (decking) with impact-resistance requirement.
Moisture resistant (*Type III*)	– improved resistance to the effects of moisture.
Moisture resistant/ flooring (*Type II/III*)	– combines strength with resistance to impact and moisture.

Wood–cement particleboard   A mixture of wood particles and cement, producing a high-density board. Used where fire and weather resistance are required, for example.

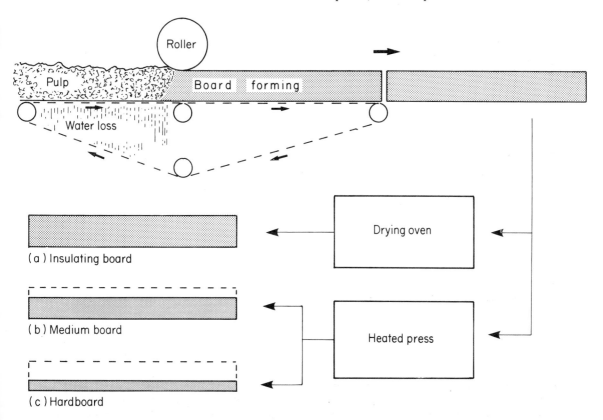

Fig. 1.40 Fibre-building board and its manufacturing processes (wet processes)

Flakeboards

Waferboard   Wafers of wood, approximately 75 mm × 75 mm and 4–6 mm thick, randomly arranged and glued together.

OSB (oriented strand board)   Narrower particles than waferboard, and more or less aligned. Boards are usually assembled in three layers – surface and core layers are oriented approximately at right angles.

Fibre board (Fig. 1.40)

Fibre building boards are produced from wood which has been shredded into a fibrous state and then reassembled into a uniform sheet form.

   The wood is first broken down into small chips then steam-treated to soften the lignin (natural resin) which binds the fibres together. Water is added (wet process) to produce a wood pulp which is spread on to a slow-moving board-forming machine (see Fig. 1.40), where it is rolled out to a uniform thickness (most of the water has been removed before this operation). What follows will depend on the required density of the finished board.

   Softboard is a low density non-compressed board which has been dried in an oven. Its cellular core is the result of moisture having evaporated during its drying process and gives the board good insulation properties – hence its name *insulation board*. Medium board and hardboard are compressed between heated presses to a density suitable for their use.

Insulation board – softboard (Fig. 1.40(a)) – a lightweight board available in sheet or tile form, with surfaces which are flat, grooved, patterned, or with stopped holes etc. Used on ceilings, walls, and under or within floors.

Medium board – three types of mid-density boards (between insulation board and hardboard):

1   Type LM (Fig. 1.40(b)) – low-density board. Used for display and notice boards etc.
    – drawing pins can easily enter the board.
2   Type HM (Fig. 1.40(b)) – panel board
    – higher density than LM. Used for wall lining, partitions, etc.
3   Type MDF (medium-density fibreboard) – made by a 'dry' process with resin binders. It has two smooth faces – without the mesh textured surface usually associated with boards which have been manufactured using the wet

process. Used for furniture, decorative mouldings, frames for flush doors, etc.

Hardboard (Fig. 1.40(c)) – a high-density sheet material with one smooth face. An ideal panel and lining material, often used as a cheaper alternative to plywood. Available with an enamelled or lacquered surface, or with a patterned or textured finish. Perforated boards (pegboard) can be used for display or as a means of ventilation.

Tempered hardboard – standard hardboard impregnated with oil or resin to increase its strength and water resistance, making it suitable for exterior use.

1.10 Laminated plastics

Laminated plastics are thin synthetic (man-made) plastics veneers and are capable of providing a both decorative and hygienic finish to most horizontal and vertical surfaces.

   Figure 1.41 shows how the laminations are built-up before being bonded together by a combination of heat and pressure.

   The finished sheets are usually supplied in large sheet sizes, but, because of their thickness (not usually more than 1.5 mm) and their hard surface, handling can prove difficult, for they are liable to split or shatter if they are subjected to sharp or sudden bends or blows, some types being more brittle than others.

   Because of their flimsy nature, these sheets are best stored flat (face to face), otherwise they will be inclined to take on and retain the shape that they are left in, i.e. bent, bowed, twisted, etc. A further precaution when storing is to ensure that no grit is trapped between the sheets' decorative surfaces, otherwise, as one sheet is drawn over the other, the chances are that its surface will be permanently scratched.

Cutting and trimming

The veneers can be cut successfully by using hand and/or machine tools.

Hand tools   Cutting can be carried out by using either a sharp fine-toothed tenon saw, cutting from the decorative face, or by scoring through the decorative face with a purpose-made scoring tool then gently lifting the waste or off-cut side, thus closing the V and allowing the sheet to break along the scored line.

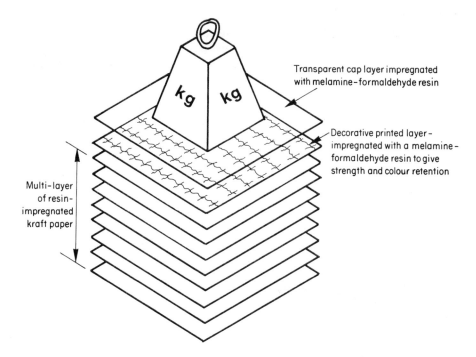

Transparent cap layer impregnated with melamine-formaldehyde resin

Decorative printed layer – impregnated with a melamine-formaldehyde resin to give strength and colour retention

Multi-layer of resin-impregnated kraft paper

Fig. 1.41 Laminated plastic the composition

A block plane and/or file can be used to trim edges. Always remember to keep your hands and fingers away from these edges while this process is being carried out – processed edges can be very sharp.

Machine tools Because of its hard and brittle nature, special care must be taken both with regard to the method of holding laminated-plastics material while it is being cut and during the machining process.

Special blades and cutters (tungsten carbide-tipped, 'TCT') are available and are advisable for processing this material. While processing operations are being carried out there is always the risk of injury to the eyes. It is therefore essential that eye protection be worn, not only by the operator but also by others in close proximity to the operation.

Veneer application (Fig. 1.42)

Because laminated plastics are used as veneers, their application on to board materials should be dealt with in a similar manner to that of a wood veneer; that is to say, if the backing board is to retain its shape, i.e. flatness, it must be kept in balance, therefore any veneer or additional veneers

(in the case of plywood) applied to one face (Fig. 1.42(a)) should have an equivalent compensating veneer applied to the opposite face (Fig. 1.42(b)). In the case of laminated plastics, balancing or counter veneers are usually without a decorative finish, making them less expensive. An alternative to using a counter veneer is to secure the whole of the underside of the board to an under-frame or sub-frame (Fig. 1.42(c)).

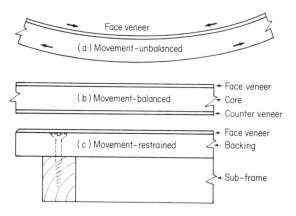

Face veneer
(a) Movement – unbalanced

(b) Movement – balanced
Face veneer
Core
Counter veneer

(c) Movement – restrained
Face veneer
Backing
Sub-frame

Fig. 1.42 Veneering

1.11 Board and sheet sizes

Table 1.6 gives some indication of the many board and sheet sizes available at timber merchants. It may be found, however, that future sheet sizes could follow a pattern of co-ordinating a few standard sizes. For example;

Width (mm): 600, 900, 1200
Length (mm): 1800, 2400, 2700, 3000

Table 1.6　Manufactured board and sheet-material sizes

Sheet material	Length(s) (mm)	Width(s) (mm)	Thickness (mm)
Plywood	Combination of the following sizes: 3050, 2745, 2440, 2135, 1830, 1525, 1270, 1220, 915 × 1830, 1525, 1270, 1220, 915		3-30
Laminboard and blockboard	1220 1830 2440	1220 2440	12-38
Particle board (chipboard)	2440 3660 4575	610 1220	2-50
Hardboard	1220, 1830, 2440, 3660, up to 5490	610 1220	2-12
Medium board	1830-3660	1220	6-22
Insulation board	1830 2440 3050 3660	1220	9-25
Laminated plastics	2135 2440 3050 3660	915 1220 1525	0.8-1.5

Sample sheet size: say 1200 mm × 2400 mm or 1200 mm × 3000 mm

and so on. Note: when ordering plywood, the first dimension given will be the length parallel to the grain of the face veneer. For example, a sheet 2440 × 1220 would have the grain of its face veneer running parallel to its longest edge – therefore known as a *long grain board*. A *short grain board* would have its face grain parallel to the shortest dimension.

1.12 Adhesives

Adhesives are made from either natural or synthetic materials. Table 1.7 lists six of these adhesives, together with some of their general characteristics. The following notes briefly describe these adhesives and some common terminology associated with them.

Adhesive type

Casein　Derived from soured milk curds, which are dried, treated, and mixed with chemicals to produce a powder which, when mixed with water, is ready for use. It is used in general joinery assembly work and in the manufacture of plywood. It tends to stain some woods.

Urea formaldehyde (UF)　A synthetic-resin adhesive, chemically cured (hardened) and available as a single- or two-part (two-component) adhesive. It is used for assembly work, veneering, and in the manufacture of plywood and particle board.

Table 1.7　Adhesive characteristics (general guide only as properties may differ)

Adhesive type and classification		Moisture resistance	Form and components									General use
			Single		Double		Gap filling			Bond pressure		
			PD	LQ	PD/LQ	LQ/LQ	Yes	MB	No	L	H	
N	Casein	–	Poor	√					√		√	√	Assembly work, plywood mnf.
SR	Urea formaldehyde (UF)	+ MR	Fair	√		√	√	√	√		√	√	Assembly work, veneering, plywood mnf., particle-board mnf.
SR	Resorcinol formaldehyde (RF)	WBP	Good			√	√	√			√		Outdoor timber structures
SR	Phenol formaldehyde (PF)	WBP	Good			√			√			√	Plywood mnf., particle-board mnf.
SR	Polyvinyl acetate (PVA)	–	†Poor	√					√		√		Assembly work, veneering
N&S	Contact	–	Fair	√*					√		√		Veneering, laminated plastics

* Available in 'gel' form
† Could be better, depending on type
Key: N - natural　SR - synthetic resin　N&S - natural and synthetic　MR - moisture-resistant　+ MR - modified (UF) may be BR (boil resistant)　WBP - weather- and boil-proof　PD - powder　LQ - liquid　MB - maybe　L - low　H - high　mnf. - manufacture

Resorcinol formaldehyde (RF) Composition similar to UF. A two-part adhesive used for outdoor timber structures.

Phenol formaldehyde (PF) Components similar to RF. Used in the manufacture of plywood and particle board.

Polyvinyl acetate (PVA) A thermoplastic adhesive, cured mainly by evaporation. PVA is a simple-to-use one-part emulsion type of adhesive, used extensively for glueing joinery components and veneering. This type of adhesive has been responsible for the dramatic decline in use of the once most popular joinery adhesive of all – animal glue, which was derived from animal bones and skins.

Contact adhesives Adhesives made of natural or synthetic rubber and a solvent which evaporates when exposed to the air, giving off a heavy flammable vapour. These adhesives are used in the application of laminated plastics – bonding is achieved by coating both surfaces to be joined, leaving them to become tacky (for a time specified by the manufacturer), then laying one on to the other without trapping any air under the surfaces. Bonding is instantaneous on contact (hence the name *contact* adhesive), with the exception of *thixotropic* types which allow a certain amount of movement for minor adjustments.

Adhesive characteristics

Form Adhesives may be of the one- or two-component types – liquid, powder, or both. Two-component types become usable either by applying them direct from their containers, mixing their components together, or by applying each of their parts separately to the surfaces being joined. Some types, however, simply have to be mixed with water.

Moisture resistance This refers to the adhesive's inherent ability to resist decomposition by moisture. Resistance is classified in the following manner;

MR	–	moisture (and moderately weather) resistant,
BR	–	boil resistant,
WBP	–	weather and boil-proof,

to which reference has been made in Section 1.9 with regard to the bonding of plywood veneers.

Gap filling Adhesives which qualify as gap-filling adhesives should be capable of spanning a 1.3 mm gap without crazing. They are used in situations where a tight fit cannot be assured.

Bond pressure This refers to the pressure necessary to ensure a suitable bond between the two or more surfaces joined together. The means by which pressure is applied is usually by either mechanical or manual presses or by clamps of various shapes and sizes. Wood wedges can be used not only to apply pressure but also to retain it permanently. The length of time needed to secure a bond will vary with each type of adhesive, its condition, and the surrounding temperature.

Storage life (may be referred to as *shelf life*) This is the stated time that the containerised adhesive will remain stable – beyond this period, marked deterioration may occur, affecting the strength and setting qualities.

Shelf life Once the adhesive's components have been exposed to the atmosphere, the storage life will usually be shortened. Shelf life therefore may or may not refer to the usable period after the initial opening – always note manufacturers' recommendations.

Pot life This is the length of time allowed for the adhesive to start to harden after either mixing or preparing the adhesive for use. (Note: working or assembly times can be taken as the time needed for hardening once the adhesive has been applied or spread on to the workpiece.)

Application of adhesives

Methods and equipment used in applying adhesives will depend on the following factors:

1. type of adhesive,
2. width of surface to be covered,
3. total surface area,
4. work situation,
5. clamping facilities,
6. quantity of work.

The spreading equipment in question could therefore be any one of the devices listed below:

a) mechanical spreader,
b) roller,
c) brush,
d) spatula,
e) toothed scraper.

Caution

All forms of adhesives should be regarded as being potentially hazardous if they are not used in accordance with the manufacturer's instructions – either displayed on the container or issued as a separate information sheet. Depending on the type of adhesive being used, failure to carry out the precautions thought necessary by the manufacturer could result in

a) an explosion – due to the adhesive's flammable nature or flammable vapour given off by it;

b) poisoning – due to inhaling toxic fumes or powdered components;

c) skin disorders (dermatitis) – due to contact while mixing or handling uncured adhesives. *Always cover skin abrasions before starting work.*

Where there is a risk of dermatitis, use a barrier cream or disposable protective gloves.

Always wash hands thoroughly with soap and water at the end of a working period.

2

Hand tools and workshop procedures

2.1 Measuring tools

Measuring tools are used either to transfer measurements from one item to another or for checking pre-stated sizes.

Scale rule

At some stage in your career you will have to take sizes from or enter sizes on to a drawing – you must therefore familiarise yourself with methods of enlarging or reducing measurements accordingly. It is essential to remember that all sizes stated and labelled on working drawings will be true full sizes, but for practical reasons these sizes will in some cases have to be proportionally reduced to suit various paper sizes, by using one of the following scales: 1:2 (half full size), 1:5, 1:10, 1:20, 1:50, 1:100, 1:200, 1:1250, 1:2500.

Figure 2.1 illustrates the use of a scale rule, which enables lengths measured on a drawing to be converted to full-size measurements and vice versa.

Four-fold metre rule

This rule should have top priority on your list of tools. Not only is it capable of accurate measurement, it is also very adaptable (see Fig. 2.2). It is available in both plastics and wood, and calibrated in both imperial and metric units. Some models (clinometer rules) also incorporate in their design a spirit-level and a circle of degrees from 0 to 180°.

With care, these rules will last for many years. It is therefore important when choosing one to find the type and make that suits your hand. Ideally, the rule should be kept about your person while at work. The most suitable place while working at the bench, or on site at ground level is usually in a rule slide pocket sewn to the trouser leg of a bib and brace or overall etc. The use of a seat or back pocket is not a good idea.

Flexible steel tapes (Fig. 2.3)

These tapes retract on to a small enclosed spring-loaded drum and are pulled out and either pushed back in or have an automatic return which can be stopped at any distance within the limit of the tape's length. Their overall length can vary from 2 m to 5 m, and they usually remain semi-rigid for about the first 500 mm of their length. This type of tool is an invaluable asset, particularly when involved in site work, as it fits easily in the pocket or clips over the belt.

It seems to have become common practice of late to use a tape as a substitute for a metre folding rule, though it is better used to complement it.

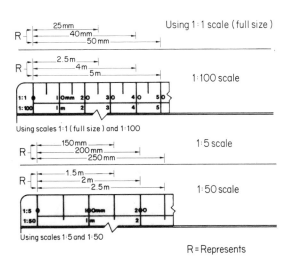

Fig. 2.1 Using a scale rule

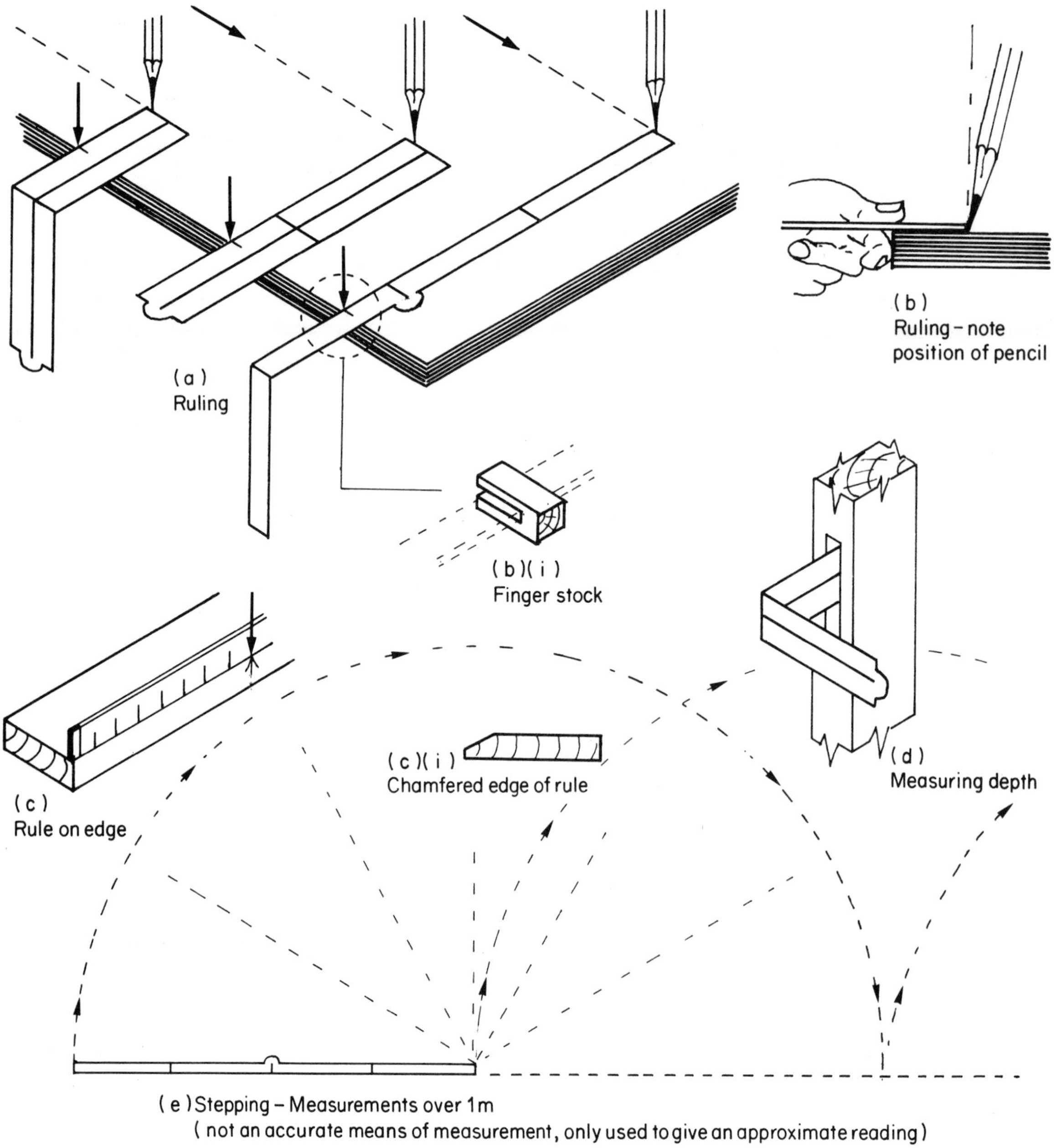

(a)
Ruling

(b)
Ruling – note
position of pencil

(b)(i)
Finger stock

(c)
Rule on edge

(c)(i)
Chamfered edge of rule

(d)
Measuring depth

(e) Stepping – Measurements over 1m
(not an accurate means of measurement, only used to give an approximate reading)

Fig. 2.2   The versatility of a four-fold metre rule
a)   Ruling different widths
b)   Standard method of holding rule at (a)
b)(i)   Ruling aid - prevents cut fingers and splinters
c)   Transferring measurements
c)(i)   Chamfered edge of rule - provide accurate flat reading
d)   Rule used as a depth gauge
e)   Stepping - measurements over 1 m (not an accurate
     means of measurement, only used to give an approximate
     reading)

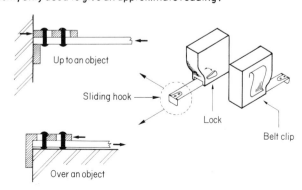

Up to an object

Sliding hook

Lock

Belt clip

Over an object

Fig. 2.3   Flexible steel tapes

2.2 Setting-out and marking-out (marking-off) tools

The drawings produced by the designer of a piece of work are usually reduced to an appropriate scale (Fig. 2.1) so that an overall picture may be presented to the client. Once approval has been given, the setting-out programme can begin. This will involve redrawing various full-size sections through all the components necessary for the construction, to enable the joiner to visualise all the joint details etc. and make any adjustments to section sizes.

Setting-out is done on what is known as a *rod*. A rod may be a sheet of paper, hardboard, or plywood, or a board of timber. By adopting a standard setting-out procedure, it is possible to simplify this process. For example (see Fig. 2.4):

1 draw all sections with their face side towards you (Fig. 2.4(a)),
2 draw vertical sections (VS) first – with their tops to your left (Fig. 2.4(b)),
3 draw horizontal sections (HS) above VS – keeping members with identical sections in line on HS and VS, e.g. top rail with stile in Fig. 2.4(c),
4 allow a minimum of 20 mm between sections (Fig. 2.4(d));
5 dimension only overall heights, widths, and depths.

Fig. 2.5 Using dividers or compasses

Setting-out will involve the use of some, if not all, of the following tools and equipment;

a) a scale rule (Fig. 2.1),
b) a straight-edge,
c) a four-fold metre rule (Fig. 2.2),
d) drafting tape, drawing-board clips, or drawing pins,

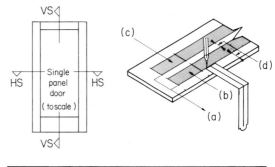

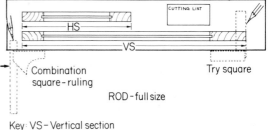

Fig. 2.4 Setting-out a rod

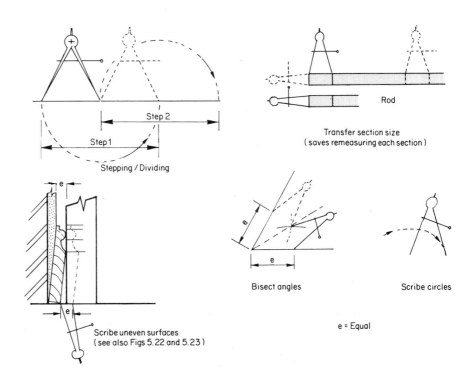

e)   an HB pencil,
f)   a try-square (Figs 2.4 and 2.9),
g)   a combination square (Figs 2.4 and 2.9),
h)   dividers (Fig. 2.5),
i)   compasses (Fig. 2.5),
j)   a trammel (Fig. 2.6).

Figure 2.4 shows a space left on the rod for a cutting list. This is an itemised list of all the material sizes required to complete a piece of work. A typical cutting list is shown in Fig. 2.7, together

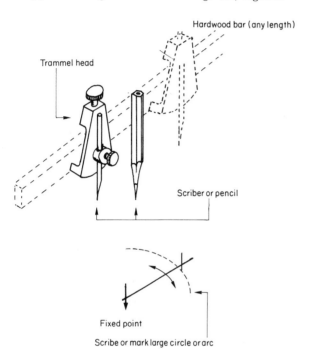

Fig. 2.6   Trammel

Job Title						Job No.
Quantity		Saw Size (Ex.)			Finish Size	
No.	Item	L	W	T	W	T	Remarks

Sundries – Nails, Screws, Hardware, etc.	
Quantity	Description

L = Length, W = Width, T = Thickness

Fig. 2.7   Cutting and ironmongery list

with provision for items of hardware (ironmongery) etc. It follows that the information with regard to timber sizes and quantity will be required by the wood machinist (Chapter 4).

Marking-out or marking-off involves the transfer of rod dimensions on to the pieces of timber and/or other materials needed. Provided that the rod is correct (double check), its use (see Fig. 2.8) reduces the risk of duplicating errors, especially when more than one item is required.

Once all the material has been reduced to size (as per the cutting list) and checked to see that its face side is not twisted and that all the face edges are square with their respective face sides, the marking-out process can begin.

Figure 2.8 illustrates a typical marking-out procedure for a simple mortise-and-tenoned frame.

Marking-out (marking-off) tools

Try-squares (Fig. 2.9)   As their name suggests, these test pieces of timber for squareness or are used for marking lines at right angles from either a face side or a face edge.

It is advisable periodically to test the try-square for squareness (see Fig. 2.9(a)). Misalignment could be due to misuse or accidentally dropping it on to the floor.

Large all-steel graduated try-squares are available without a stock which makes them very useful when setting-out on large flat surfaces. They look like the steel roofing square shown in *Carpentry and Joinery 2*.

Combination square (Fig. 2.9(b))   This can be used as a try-square, but has the added advantage of being very versatile, in that it has many other uses, e.g. as a mitre square (marking and testing angles of 45°), height gauge, depth gauge, marking gauge (see Fig. 2.4), spirit-level (some models only), and rule.

It is common practice to use a pencil with a square, as shown in Fig. 2.9(c), although a marking knife (Fig. 2.9(d)) is sometimes used in its place (especially when working with hardwoods) to cut across the first few layers of fibres so that the saw cut which follows leaves a sharp clean edge, e.g. at the shoulder line of a tenon.

Marking and mortise gauges (Fig. 2.10)   As can be seen from the diagram, these gauges are similar in appearance and function, i.e. scoring lines parallel to the edge of a piece of timber. The main difference is that the marking gauge scores only a single line but the mortise gauge scores two in one

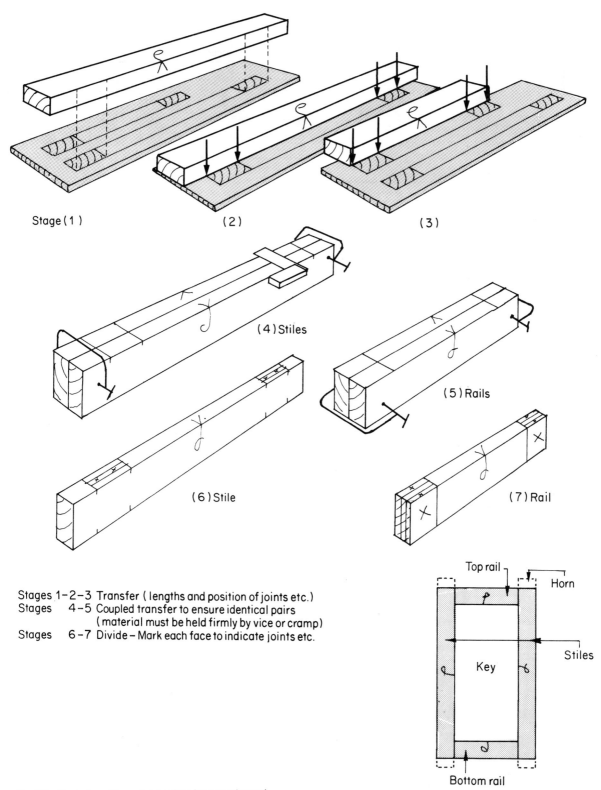

Stage (1)   (2)   (3)

(4) Stiles

(5) Rails

(6) Stile

(7) Rail

Stages 1-2-3 Transfer (lengths and position of joints etc.)
Stages   4-5 Coupled transfer to ensure identical pairs
              (material must be held firmly by vice or cramp)
Stages   6-7 Divide – Mark each face to indicate joints etc.

Top rail

Horn

Key

Stiles

Bottom rail

Fig. 2.8 Typical marking-out procedure for a mortise-and-tenoned frame

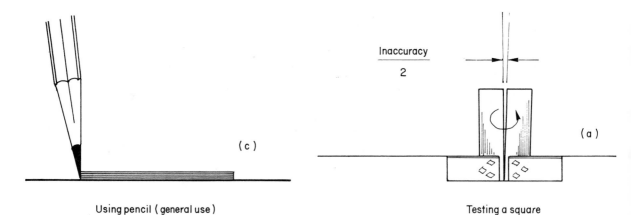

Using pencil (general use)   Testing a square

Inaccuracy
___
2

(a)

(c)

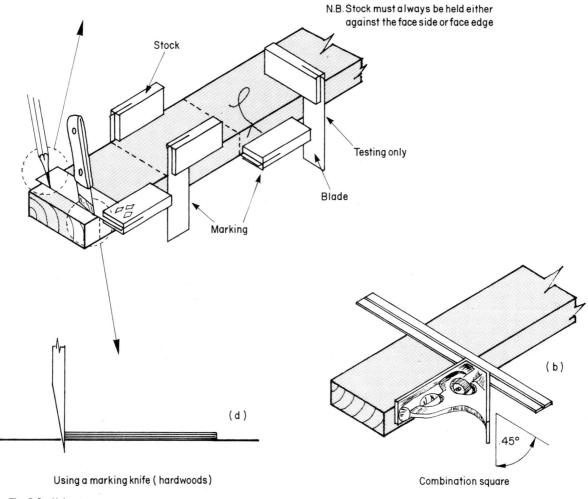

Stock

N.B. Stock must always be held either
against the face side or face edge

Testing only

Blade

Marking

Using a marking knife (hardwoods)

Combination square

45°

(b)

(d)

Fig. 2.9   Using try-squares

pass. It is possible to buy a gauge that can perform both operations simply by being turned over (see Fig. 2.10(b)).

Cutting gauge (Fig. 2.10(d)) This is used to cut across the fibres of timber. It therefore has a similar function to that of a marking knife.

Marking angles and bevels

The combination square (previously mentioned, and shown in Fig. 2.9) has probably superseded the original mitre square, which looked like a set square but had its blade fixed at 45° and 135° instead of 90°. However two of the most useful pieces of bench equipment when dealing with mitres are a mitre template and a square and mitre template. Examples of their use are shown in Fig. 2.11.

Any angles other than 45° will have to be transferred with the aid of either a template pre-marked from the rod or the site situation, or by using a sliding bevel. This has a blade which can slide within the stock and be locked to any angle (Fig. 2.12). The bevel as a whole can sometimes prove to be a little cumbersome for marking dovetail joints on narrow boards. This can be overcome quite easily by using a purpose-made dovetail template (see Fig. 2.12).

2.3 Saws

Saws are designed to cut both along and across the grain of wood (except the rip saw – see Table 2.1), and the saw's efficiency will be determined by

1 the type and choice of saw,
2 the saw's condition,
3 the application,
4 the material being cut.

Square over moulded section

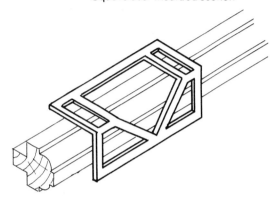

SQUARE AND MITRE TEMPLATE

Score mitre profile with chisel or marking knife held flat

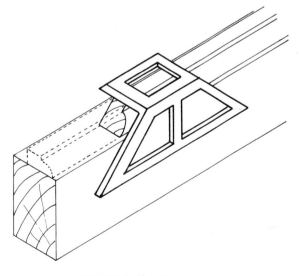

MITRE TEMPLATE
N.B. Not intended as a chisel guide

Fig. 2.11 Marking moulded sections before cutting

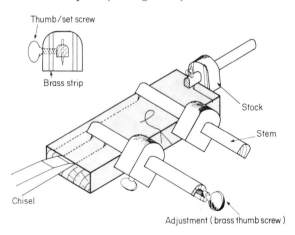

Thumb / set screw
Brass strip
Stock
Stem
Chisel
Adjustment (brass thumb screw)

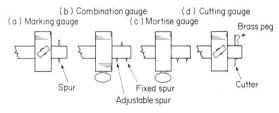

(a) Marking gauge
(b) Combination gauge
(c) Mortise gauge
(d) Cutting gauge
Brass peg
Spur
Fixed spur
Adjustable spur
Cutter

Fig. 2.10 Using marking and mortise gauges

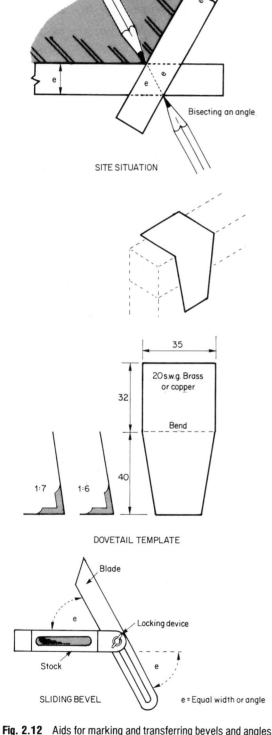

SITE SITUATION

Bisecting an angle.

DOVETAIL TEMPLATE

35

20 s.w.g. Brass
or copper

32

Bend

40

1:7   1:6

SLIDING BEVEL

Blade

Locking device

Stock

e

e

e = Equal width or angle

Fig. 2.12   Aids for marking and transferring bevels and angles

Choice of saw

Broadly speaking, saws can be categorised into
four groups;

a)   handsaws,
b)   backed saws,
c)   framed saws,
d)   narrow-blade saws. } for cutting curves

As can be seen from Table 2.1, each type/group
can be further broken down into two or three
specifically named saws, which are available in a
variety of sizes and shapes to suit particular
functions.

Condition of saw

It is important that saws are kept clean (free from
rust) and sharp at all times (see Section 2.12). Dull
or blunt teeth not only reduce the efficiency of the
saw, but also render it potentially dangerous. For
example, insufficient set could cause the saw to jam
in its own kerf and then buckle, or even break (see
Fig. 2.23).

Method of use

The way the saw is used will depend on the
following factors;

a)   the type and condition of the wood being cut,
b)   the direction of cut – ripping or cross-cutting,
c)   the location – bench work or site work.

Practical illustrated examples are shown in
Figs 2.13 to 2.22. Note the emphasis on safety, i.e.
the position of hands and blades and body balance
etc.

Material being cut

A vast variety of wood species are used in the
building industry today, and many, if not all, will
at some time be sawn by hand. The modern saw is
ideally suited to meet most of the demands made
upon it, although there are instances where it will
be necessary to modify general sawing techniques,
for example when dealing with wood that is

a)   very hard,
b)   of very high moisture content,
c)   extremely resinous,
d)   case-hardened.

Occurrence of the above will depend on the type
of work being undertaken, and the measures to be
adopted will be discussed in *Carpentry and Joinery 2.*

Table 2.1 Saw fact sheet

Type or group	Saw	Function	Blade length (mm)	Teeth shape	Teeth per 25 mm	Handle type	Remarks
Handsaws	Rip (Fig. 2.13)	Cutting wood with the grain (ripping)	650		4–6		Seldom used below 6 teeth per 25 mm
	Cross-cut (Fig. 2.14)	Cutting wood across the grain	600 to 650		7–8		Can also be used for rip sawing
	Panel (Figs 2.15 and 2.16)	Cross-cutting thinner wood and manufactured board (M/B)	500 to 550		10		Easy to use and handle
Backed saws	Tenon (Figs 2.17 and 2.18)	Tenons and general bench work	300 to 450		12–14		Depth of cut restricted by back strip (blade stiffener)
	Dovetail (Fig. 2.19)	Cutting dovetails and fine work	200 to 250		18–20		
Framed saws	Bow	Cutting curves in heavy sectioned timber and M/B	200 to 300		12 + or −		Radius of cut restricted by blade width
	Coping (Fig. 2.20)	Cutting curves in timber and M/B	160		14		Thin narrow blade
	Hacksaw (Fig. 2.21)	Cutting hard and soft metals	250 and 300		14–32		Small teeth - to cut thin materials. The larger the teeth, the less liable to clog - small frame hacksaw (see Fig. 2.23)
Narrow-blade saws	Compass	Cutting slow curves in heavy and large work	300 to 450		10 + or −		Interchangeable blades of various widths - unrestricted by a frame
	Pad or key hole	Enclosed cuts - piercing panels, etc.	200 to 300		10 + or −		Narrow blade partly housed in handle, therefore length adjustable

Key: M/B-manufactured board

Fig. 2.13 Ripping with a handsaw

Fig. 2.14 Cross-cutting with a handsaw

Fig. 2.15   Sawing down the grain (vertically) with a panel saw

Fig. 2.17(a)   Tenon saw starting a cut: vice held

Fig. 2.16   Sawing down the grain - material angled, two saw lines are visible

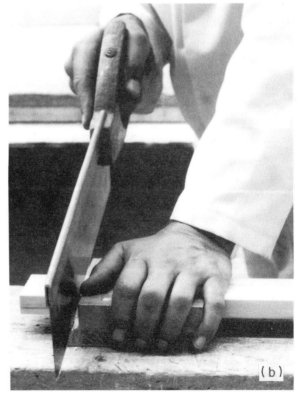

Fig. 2.17(b)   Tenon saw starting a cut: using a bench hook

Fig. 2.18 Sawing down the grain - tenon saw

Fig. 2.19 Dovetail saw - starting a cut

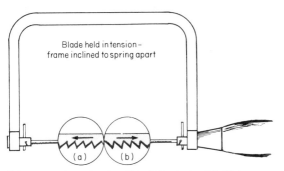

Blade held in tension -
frame inclined to spring apart

(a) To cut on a forward stroke will tend to bend or break the blade
(b) To cut on a back stroke will tend to keep blade taut

Fig. 2.20 Using a coping saw

Fig. 2.21 Using a hacksaw

Fig. 2.22 Using a small hacksaw

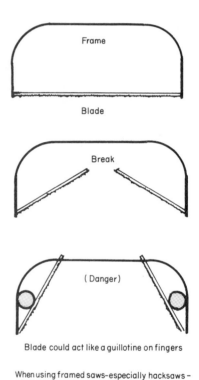

Frame

Blade

Break

(Danger)

Blade could act like a guillotine on fingers

When using framed saws-especially hacksaws -
always keep fingers outside the framework
in case the blade breaks

Fig. 2.24 Frame-saw safety

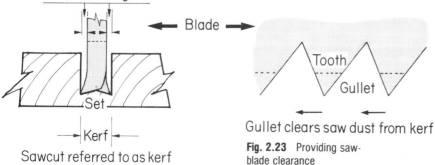

Saw set provides clearance - prevents
saw from binding in kerf

Blade

Set

Kerf

Sawcut referred to as kerf

Tooth

Gullet

Gullet clears saw dust from kerf

Fig. 2.23 Providing saw-
blade clearance

2.4 Planes

There are many types of plane. All are capable of cutting wood by producing shavings, but not all are designed to produce *plane* flat surfaces as the name implies. However, as can be seen from Table 2.2, each plane has its own function and, for the sake of convenience, planes have been placed within one of two groups;

a) bench planes,
b) special purpose planes.

Bench planes (Figs 2.25 – 2.28)

Wooden-bodied bench planes have been superseded by the all-metal (with the exception of the handle and front knob) plane, although the wooden jack plane is still regarded by some joiners as the ideal site plane, as it is light to handle and less liable to break if dropped. Probably the greatest asset of the wooden bench plane is its ability to remove waste wood rapidly – where accuracy is not too important.

All metal bench planes are similarly constructed with regard to blade angle (45°), adjustment, and alignment (Fig. 2.39); variations are primarily due to the size or design of the plane sole, which determines the function (Table 2.2).

Figures 2.40 and 2.41 show a jack plane being used for flatting and edging a short piece of timber. Notice particularly the position of the hands in relation to the operation being carried out.

Table 2.2 Plane fact sheet

Group	Plane	Function	Length (mm)	Blade width (mm)	Remarks
Bench planes	Smoothing (Fig. 2.25(a))	Finishing flat surfaces	*240, *245, 260	45, 50, 60	50 mm the most common blade width.
	Record CS88 (Fig. 2.25(b))	As above	245	60	'Norris' type cutter adjustment - combining depth of cut with blade alignment.
	Jack (Fig. 2.26)	Processing sawn timber	*355, 380 mm	50, 60	60 mm the most common blade width.
	Fore Jointer (Fig. 2.27) (try plane)	Planing long edges (not wider than the plane's sole) straight and true	*455 *560, 610	60 60, 70	The longer the sole, the greater the degree of accuracy.
	Bench rebate (carriage or badger) (Fig. 2.28)	Finishing large rebates	235, 330	54	Its blade is exposed across full width of sole.
		* Available with corrugated soles (Fig. 2.29) which are better when planing resinous timber.			
Special planes	Block (Fig. 2.30)	Trimming - end grain	140, 180, 205	42	Cutter seats at 20° or 12°(suitable for trimming laminated plastics), depending on type.
	Circular (compass plane) (Fig. 2.31)	Planing convex or concave surfaces	235, 330	54	Spring-steel sole adjusts from flat to either concave or convex.
	Rebate (Fig. 2.32)	Cutting rebates with or across the grain	215	38	Both the width and depth of rebate are adjustable.
	Should/rebate (Fig. 2.33)	Fine cuts across grain, and general fine work	152, 204	18, 25, 29, 32	Some makes adapt to chisel planes.
	Bullnose/shoulder/rebate (Fig. 2.34)	As above, plus working into confined corners	100	25, 29	
	Side rebate (Fig. 2.35)	Widening rebates or grooves - with or across grain	140		Removable nose - works into corners. Double-bladed - right and left hand. Fitted with depth gauge and/or fence.
Special planes (continued)	Plough (Fig. 2.36)	Cutting grooves of various widths and depths - with and across grain	248	3 to 12	Both width and depth of groove are adjustable.
	Combination	As above, plus rebates, beading, tongues, etc.	254	18 cutters, various shapes and sizes	Not to be confused with a 'multi-plane', which has a range of 24 cutters.
	Open-throat router (Fig. 2.37)	Levelling bottoms of grooves, trenches, etc.		6, 12 and V	A fence attachment allows it to follow straight or round edges.
	Spokeshave (Fig. 2.38): (a) flat bottom, (b) round bottom (Fig. 2.50)	Shaving convex or concave surfaces - depending on type (a) or (b)	250	54	Available with or without 'micro' blade depth adjustment.

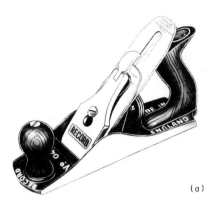

(a)

Fig. 2.25(a)   Smoothing plane

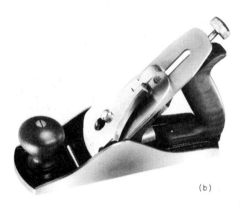

(b)

Fig. 2.25(b)   Record CS88 plane

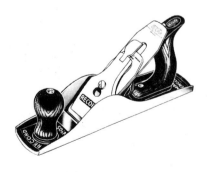

Fig. 2.26   Jack plane

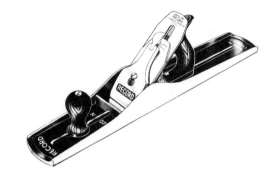

Fig. 2.27   Jointer or try-plane

Fig. 2.28   Bench rebate (carriage or badger) plane

Fig. 2.29   Corrugated sole

Fig. 2.30   Block plane

Processing a piece of sawn timber by hand is carried out as follows:

1 Select and, using a jack plane, plane a face side (best side) straight (Fig. 2.42(a)) and out of twist (Fig. 2.42(b)). (*Winding laths* are used to accentuate the degree of twist.) Label the side with a face-side mark (Fig. 2.9).

2 Plane a face edge straight and square to the face side (keep checking with a try-square), using either a jack plane or a try-plane (Fig. 2.43), depending on the length of the timber being processed. As can be seen from Fig. 2.43(a) the shorter the sole the more difficult it is to produce long straight edges. Long lengths will require end support to prevent tipping – this can be achieved by positioning a peg in one of a series of pre-bored holes in the face or leg of the bench, see Fig. 2.44. On completion, credit the edge with a face-edge mark (Fig. 2.9).

3 Gauge to width (Fig. 2.10), ensuring that the stock of the gauge is held firmly against the face edge at all times, then plane down to the gauge line.

4 Gauge to thickness – as in step 3 but this time using the face side as a guide.

Note: when preparing more than one piece of timber for the same job, each operation should be carried out on all pieces before proceeding to the next operation, i.e. face side all pieces, face edge all pieces, and so on.

The smallest and most used of all the bench planes is the smoothing plane. This is very easy to handle and, although designed as a fine finishing plane for dressing joints and surfaces alike, it is used as a general-purpose plane for both bench and site work. Figure 2.45 shows a smoothing plane being used to dress (smooth and flat) a panelled door, and how by tilting the plane it is possible to test for flatness (this applies to all bench planes) – the amount and position of the light showing under its edge will determine whether the surface is round or hollow.

Note: the direction of the plane on the turn at corners or rail junctions – for example, working from stile to rails or rail to stile – will be determined by the direction of the wood grain.

Special planes (Figs 2.30 – 2.38)

It should be noted that there are more special planes than the ten planes listed in Table 2.2, and there are also variations in both style and size.

The kinds of planes selected for your tool kit will depend on the type of work you are employed to do. There are, however, about four planes which, if not essential in your work, you should find very useful. They are

a) a block plane (Fig. 2.30),
b) a rebate plane (Fig. 2.32),
c) a plough plane (Fig. 2.36),
d) spokeshaves (Fig. 2.38).

Block plane (Fig. 2.30) This is capable of tackling the awkwardest cross-grains of both hardwood and softwood, not to mention the edges of manufactured boards and laminated plastics. Some models have the advantage of an adjustable mouth (Fig. 2.118) and/or their blade set to an

Fig. 2.31 Circular (compass) plane

Fig. 2.32 Rebate plane

extra low angle of 12°. Such a combination can increase cutting efficiency. Block planes are designed to be used both single- and double-handed.

Rebate plane (Fig. 2.32) Figure 2.46 shows a rebate plane being used to cut a rebate of

controlled size, by using a width-guide fence and a depth stop. It is, however, very important that the cutting face of the plane is held firmly and square to the face side or edge of the timber throughout the whole operation.

Because the blade of a rebate plane has to extend across the whole width of its sole, a cut finger can

Fig. 2.33 Bullnose/shoulder rebate plane

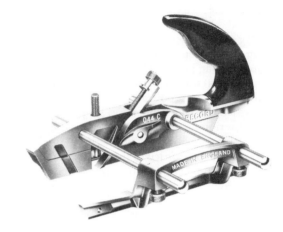

Fig. 2.36 Plough plane

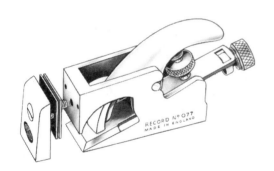

Fig. 2.34 Shoulder/rebate plane

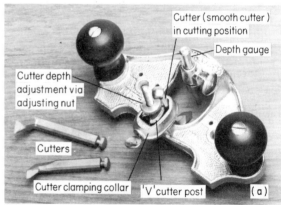

Cutter (smooth cutter) in cutting position

Depth gauge

Cutter depth adjustment via adjusting nut

Cutters

Cutter clamping collar

'V' cutter post

(a)

Fig. 2.35 Side rebate plane in use

(b)

Fig. 2.37 Open-throat router and its use

easily result from careless handling. Particular care should therefore be taken to keep fingers away from the blade during its use, and especially when

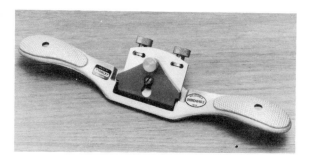

Fig. 2.38 Spoke shave

Fig. 2.40 'Flatting' (planing the surface of) a piece of timber - placed against a bench stop

Fig. 2.39 Exploded view of a Stanley bench plane
1. Cap iron screw
2. Lever cap
3. Lever screw
4. Frog - complete
5. 'Y' adjusting lever
6. Adjusting nut
7. Adjusting nut screw
8. LA (lateral adjustment) lever
9. Frog screw and washer
10. Plane handle
11. Plane knob
12. Handle screw and nut
13. Knob screw and nut
14. Handle toe screw
15. Plane bottom - sole
16. Frog clip and screw
17. Frog adjusting screw
18. Cutting iron (blade) and cap iron

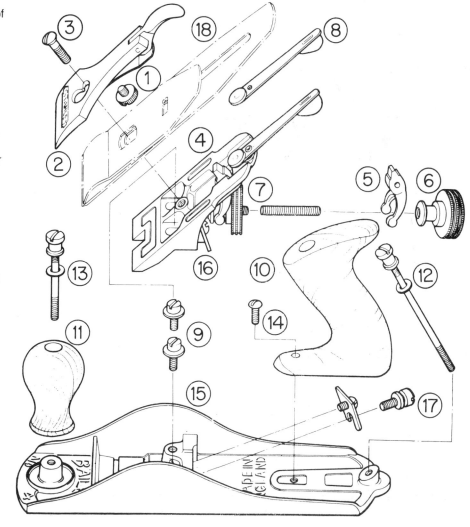

making the desired depth and/or width adjustments.

Plough plane (Fig. 2.36) Figure 2.47 shows a plough plane cutting a groove in the edge of a

Fig. 2.41 'Edging' (planing the edge of) a piece of timber - held in a bench vice with finished face positioned towards the operative - forefinger acts only as a guide to centralize the plane during the operation - not as a fence

piece of wood. It is also capable of cutting rebates to the cutter widths provided and by the method shown in Fig. 2.48.

The method of applying the plane to the wood is common to both rebate and plough planes, in that

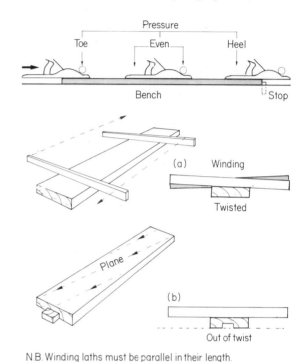

N.B. Winding laths must be parallel in their length.

Fig. 2.42 Preparing a face side

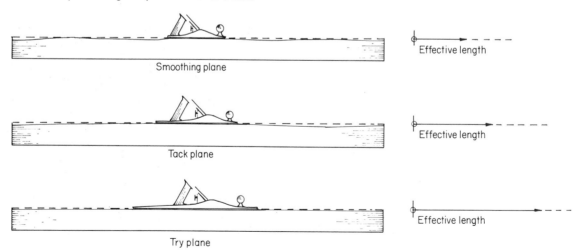

Fig. 2.43
a) Longer the plane sole - greater the accuracy for a straight edge

b) Toe down to middle

c) Even pressure (self weight of plane)

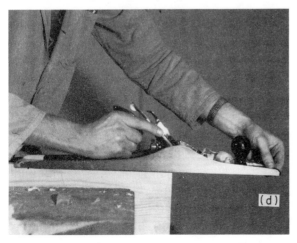

d) Heel down to end

Fig. 2.43 Using a try plane

the cut should be started at the forward end and be gradually moved back until the process is complete (see Fig. 2.49).

Joiners often prefer to use larger flat-bottomed planes for shaping convex curves, but the efficiency and ease of operation of a flat-bottomed spokeshave can only be realised when the

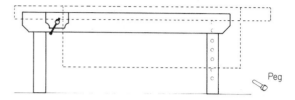

Fig. 2.44 Providing end support

Fig. 2.45 Using a smoothing plane to dress the faces of a complete (fully assembled, glued cramped and set) panelled door
a) Smoothing a door stile

b) Dressing flat a joint between door stile and rail - a slow turning action towards the joint helps to avoid unnecessary tearing of the grain

c) Smoothing a door rail

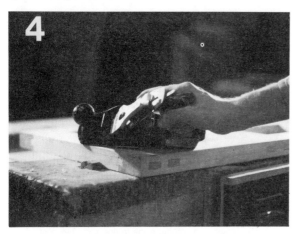

d) Testing a corner joint for flatness by tilting the plane - notice the light showing between the edge of the plane sole and the surface of the timber, indicating that at this position the surface is hollow

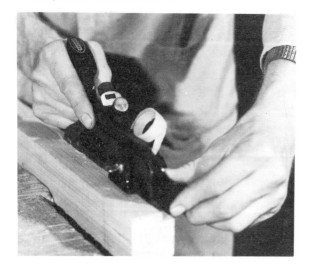

Fig. 2.46 Planing a rebate

Fig. 2.47 Plough plane cutting a groove

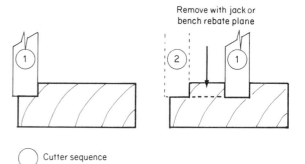

Fig. 2.48 Methods of forming a rebate with a plough plane

technique of using this tool has been mastered. Figure 2.51 shows a spokeshave in use – note particularly the position of the thumbs and forefingers, giving good control over the position of the blade in relation to its direction of cut (always with the grain).

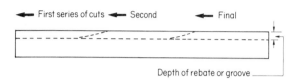

Fig. 2.49 Application of a rebate or plough plane

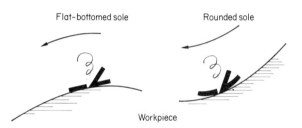

Fig. 2.50 Spokeshave - use

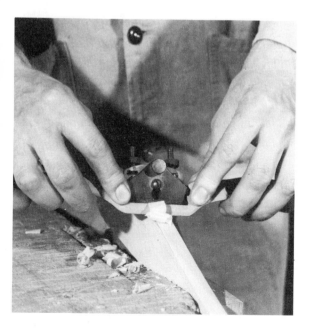

Fig. 2.51 Spokeshave being used

2.5 Boring tools

Boring tools are those tools which are capable of cutting or scraping circular holes of a predetermined·size into and below the surface of a given material. They can be divided into three groups;

1 standard bits,
2 special bits,
3 drills.

Table 2.3 has classified the above under the headings of

a) type,
b) function,
c) motive power,
d) hole size,
e) shank section.

(Table 2.3 is meant to be used in conjunction with the illustrations of tools in Figs 2.52 to 2.60.)

The driving force necessary for such tools to operate is provided by hand, by electricity (Section 4.1), or by compressed air.

Hand operations involve the use of a bit- or drill-holder (chuck) operated by a system of levers and gears. There are two main types; the *carpenter's brace* and the hand drill, also known as the *wheel-brace*.

Carpenter's brace

This has a two-jaw chuck, of either the *alligator* or the *universal* type. The alligator type has been designed to take square-tapered shanks, whereas the universal type takes round, tapered, and straight as well as square tapered shanks. The amount of force applied to the bit or drill will depend largely on the *sweep* of the brace. Figure 2.65 shows how the style and sweep can vary.

There are three main types of brace:

1 *Ratchet brace* – the ratchet mechanism allows the brace to be used where full sweeps are restricted (an example is shown in Fig. 2.66). It also provides extra turning power (by eliminating overhand movement), so often needed when boring large-diameter holes or using a turn-screw bit.

2 *Plain brace* (non-ratchet type) – limited to use in unrestricted situations only, so not recommended.

Table 2.3 Characteristics of bits and drills in common use. (Not all types are available in metric sizes.)

Group	Type of bit/drill	Function	Motive power	Range of common hole sizes	Shank section	Remarks
Bits (standard)	Centre bit (Fig. 2.52)	Cutting shallow holes in wood*	CB	$\frac{1}{4}''$ - $2\frac{1}{4}''$	□	
	Irwin-pattern solid-centre auger bit (Fig. 2.53)	Boring straight holes in wood*	CB	$\frac{1}{4}''$ - $1\frac{1}{2}''$ 6 mm–38 mm	□	General-purpose bit
	Jennings-pattern auger bit (Fig. 2.54)	Boring straight, accurate, smooth holes in wood*	CB	$\frac{1}{4}''$ - $1\frac{1}{2}''$	□	
	Jennings-pattern dowel bit (Fig. 2.55)	As above, only shorter	CB	$\frac{3}{8}''$ and $\frac{1}{2}''$ only	□	Used in conjunction with wood dowel
	Countersink (Fig. 2.56)	Enlarging sides of holes to 45°, to receive a screw head	CB HD ED	$\frac{3}{8}''$, $\frac{1}{2}''$, $\frac{5}{8}''$	□○	Rose, shell (snail) heads available depending on material being cut and speed
Bits (special)	Expansive (expansion) bit (Fig. 2.57)	Cutting large shallow holes in wood*	CB	$\frac{7}{8}''$ - 3″	□	Adjustable to any diameter within its range
	Scotch eyed auger bit (hand) (Fig. 2.58)	Boring deep holes in wood	T	$\frac{1}{4}''$ - 1″		Turned via 'T' bar. Length determined by diameter of auger.
	Forstner bit (Fig. 2.59)	Cutting shallow flat-bottomed holes in wood*	CB ED	$\frac{3}{8}''$ - 2″	□○	Ideal for starting a stopped housing
	Dowel-sharpener bit	Chamfering end of dowel	CB		□	Pointed dowel - aids entry into dowel holes
	Turn-screw bit (Fig. 2.60)	Driving large screws	CB	$\frac{1}{4}''$, $\frac{3}{16}''$, $\frac{3}{8}''$, $\frac{7}{16}''$	□	Very powerful screwdriver
	Flat bit (Fig. 2.61)	Bores holes in all forms of wood very quickly and cleanly	ED	$\frac{1}{4}''$ - $1\frac{1}{2}''$ 6 mm to 38 mm	○	Deep holes may be inclined to wander. Extension shank available.
	Screw bit (Screwmate) (Fig. 2.62)	Drills pilot, clearance and countersink in one operator	ED	Screw size 1″ × 6 to $1\frac{1}{2}''$ × 10 25 mm × 6 to 38 mm × 10	○	Saves time changing different bits
	Screw sink (Fig. 2.63)	Combination counterbore all-in-one boring of screw hole and plug hole	ED	$\frac{3}{4}''$ × 6 to 2″ × 12 19 mm × 6 to 51 mm × 12	○	Depth of counterbore can be varied. Use with Stanley plug cutter
	Plug cutter (Fig. 2.64)	Cuts plugs to fill counterbored holes	ED	To suit screw ganges 6 to 12 and hole sizes $\frac{3}{8}''$ $\frac{1}{2}''$ $\frac{5}{8}''$	○	Use only with a drill fixed into a bench drill stand, and with the wood cramped down firmly (see manufacturer's instructions for maximum safe speed).
Drills	Twist drills	Boring wood*, metals, plastics	HD ED CA	$\frac{1}{16}''$ - $\frac{1}{2}''$ 1 mm to 13 mm	○	A few sizes available as □
	Masonry drills (tungsten-carbide tipped)	Boring masonry, brickwork, concrete	HD ED	No. 6 to no. 20 $\frac{5}{16}''$ to $\frac{3}{8}''$ †	○	Available for both rotary and impact (percussion-action) drills

† Larger diameter drills are available
Key: CB - carpenter's brace   HD – hand drill (wheel-brace)   ED - electric drill (N.B. Bits used in ED must never have a screw point)   CA - compressed-air drill   □ - square tapered shank   ○ - straight rounded shank   * - wood and all wood-based products

3 *Joist brace*   –   for use in awkward spaces,
  (upright brace)   between joists for example.

Method of use   Probably the most difficult part of the whole process of boring a hole is keeping the brace either vertical or horizontal to the workpiece throughout the whole operation. Accuracy depends on all-round vision; so, until you have mastered the art of accurately assessing horizontality and verticality, seek assistance.

Fig. 2.52 Centre bit

Fig. 2.53 Irwin-pattern solid-centre bit

Fig. 2.54 Jennings-pattern auger bit

Fig. 2.55 Jennings-pattern dowel bit

Fig. 2.56 Rosehead counter sink

Fig. 2.57 Firmgrip expansive bit

Fig. 2.58 Scotch-eyed auger bit

Fig. 2.59 Forstner bit

Fig. 2.60 Bright cabinet screwdriver bit (turnscrew)

Fig. 2.61 Flat bit

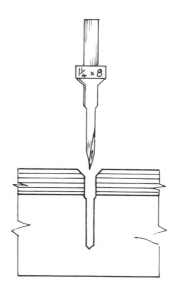

Fig. 2.62 Screw bit (Stanley Screwmate)

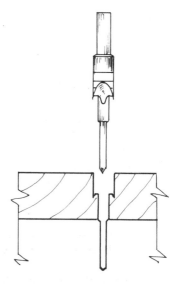

Fig. 2.63 Stanley Screwsink

Drill must be fixed into a drill stand

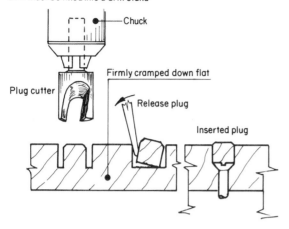

Fig. 2.64   Stanley plug cutter

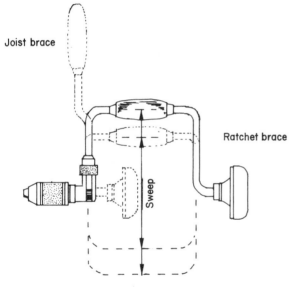

Fig. 2.65   Types of carpenter's brace

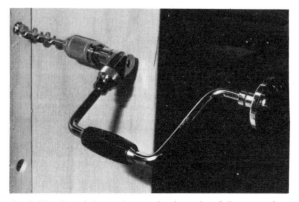

Fig. 2.66   Use of the ratchet mechanism when full sweep of the brace is restricted

Figures 2.67 and 2.68 show situations where the assistant directs the operative by simple hand signals. Note also that for vertical boring the operative's head is kept well away from the brace, giving good vision and unrestricted movement to

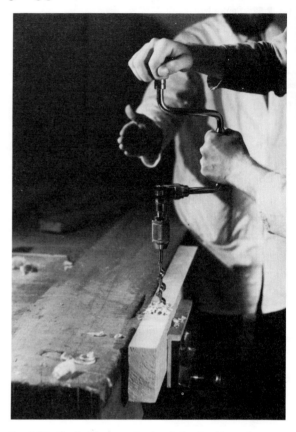

Fig. 2.67   Vertical boring using hand signal guidance from an assistant taking the vertical viewpoint

Fig. 2.68   Horizontal boring using hand signal guidance from an assistant taking the horizontal viewpoint

its sweep – only light pressure should be needed if the bit is kept sharp. Horizontal boring support is given to the brace by arm over leg as shown – in this way good balance can be achieved and maintained throughout the process. The stomach should *not* be used as a form of support.

The use of an upturned try-square etc. placed on the bench as a vertical guide is *potentially dangerous* – any sudden downward movement, due to the bit breaking through the workpiece, could result in an accident.

Figure 2.69 shows the use and limitations of some of the bits and drills mentioned, together with two methods of breaking through the opposite side of the wood without splitting it, i.e. reversing the direction of the bit or temporarily securing a piece of waste wood to the point of breaking through.

Hand drill (wheel-brace) (Fig. 2.70)

This has a three-jaw self-centring chuck, designed specifically to take straight-sided drills. It is used in conjunction with twist drills or masonry drills.

Boring devices and aids

Probably the most common of all boring devices is the bradawl (Fig. 2.71) – a steel blade fixed into a wood or plastics handle – used mainly to bore pilot holes for screw threads.

The spinal ratchet screwdriver, featured in Section 2.8, may be adapted to drill small short holes and to carry out countersinking operations.

Gauging the depth of a hole can be done either by using a proprietary purpose-made depth gauge or by making your own as in Fig. 2.72.

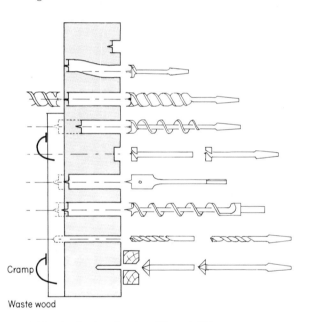

Cramp

Waste wood

Fig. 2.69 Application of various bits and drills.

Fig. 2.70 Hand drill (wheel-brace)

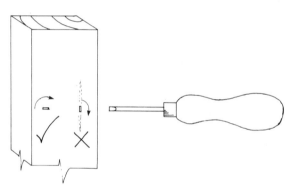

Fig. 2.71 Bradawl (pricker) - application

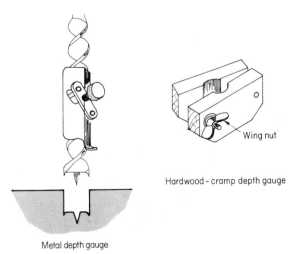

Wing nut

Hardwood - cramp depth gauge

Metal depth gauge

Fig. 2.72 Bit-depth gauge

2.6 Chisels

Wood-cutting chisels are designed to meet either general or specific cutting requirements. Table 2.4 lists those chisels in common use, together with their characteristics. Figures 2.73 to 2.78 illustrate some popular examples.

Chisels can be divided into two groups: those which cut by

a)   paring, and
b)   chopping.

Paring   This simply means the act of cutting thin slices of wood – either across end grain (Fig. 2.79) or across the grain's length (Fig. 2.80). Chisels used for this purpose are slender and designed for easy handling.

Figures 2.79 and 2.80 show two examples of paring. Note especially the method of support given to the body, workpiece, and chisel (both hands behind the cutting edge).

Chopping   These chisels are robust, to withstand being struck by a mallet. Their main function is to cut (chop) through end grain – usually to form an opening or mortise hole to receive a tenon; hence the common name *mortise chisel*.

Figure 2.81 shows a mortise chisel and mallet being used to chop out a mortise hole. Notice the firm support given to the workpiece, and how the G cramp has been laid over on its side so as not to obscure the operative's vision.

Fig. 2.73   Firmer chisel

Fig. 2.74   Bevel-edge chisel

Fig. 2.75   Registered chisel

Fig. 2.76   Mortise chisel

Fig. 2.77   Out-cannel (firmer) gouge

Fig. 2.78   In-cannel (scribing) gouge

Table 2.4   Characteristics of woodcutting chisels (not all types are available in metric sizes)

Chisel type	Function	Handle material	Blade widths	Blade section	Remarks
Firmer (Fig. 2.73)	Paring and light chopping	HW Plastics	$\frac{1}{4}''$ - $1\frac{1}{4}''$ 6 mm - 32 mm		General bench work etc. Plastics-handle types can be lightly struck.
Bevel-edged (Fig. 2.74)	As above	HW Plastics	$\frac{1}{8}''$ - $1\frac{1}{2}''$ 3 mm - 38 mm		As above, plus ability to cut into acute corners
Paring	Paring long or deep trenches	HW	$\frac{1}{4}''$ - $1\frac{1}{2}''$		Extra-long blade
Registered (Fig. 2.75)	Chopping and light mortising	HW	$\frac{1}{4}''$ - $1\frac{1}{2}''$ 6 mm - 38 mm		Steel ferrule prevents handle splitting.
Mortise (Fig. 2.76)	Chopping and heavy mortising	HW Plastics	$\frac{1}{4}''$ - $\frac{1}{2}''$ 6 mm - 13 mm		Designed for heavy impact
Gouge - firmer (out-cannel)* (Fig. 2.77)	Hollowing into the wood's surface	HW	$\frac{1}{4}''$ - $1''$ 6 mm - 25 mm		Size measured across the arc
Gouge - scribing (in-cannel)* (Fig. 2.78)	Hollowing an outside surface or edge				Extra-long blades available (paring gouge)

* Out-cannel (firmer) gouges have their cutting bevel ground on the outside; in-cannel (scribing) gouges have it on the inside.

Fig. 2.79 Vertical paring

Fig. 2.80 Horizontal paring

2.7 Shaping tools

All cutting tools can be regarded as shaping tools – just as different shapes determine the type or types of tools required to form them. The previous tools should therefore also be included under the heading of *shaping tools*.

Axe (blocker)

Provided an axe is used correctly, i.e. always keeping fingers and body behind its cutting edge, it is a highly efficient tool – invaluable to the site worker for quick removal of waste wood or for cutting wedges etc. It must be kept sharp, however, and at the finish of each operation the blade must be protected with a thick leather sheath.

Figure 2.82 shows an axe being used to cut a wedged-shaped plug (see Chapter 15). Note the piece of waste wood on the floor, to protect both the floor and the axe cutting edge.

Fig. 2.81 Chopping a mortise hole. Note the use of scrap wood to protect the bench and workpiece, and how the cramp has been laid flat with protruding bar and handle positioned towards the well of the bench, out of harm's way

Surform tools

These are a very useful versatile range of shaping tools, capable of tackling most materials, depending on the blade (Table 2.5). They are undoubtedly a valuable asset for joiners involved in house maintenance, where conventional tools are often impractical, and the bench-hand will also find these files and shapers very useful.

Figure 2.83 shows just two of these tools, and Table 2.5 shows their blade capabilities.

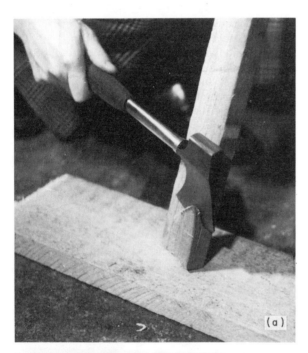

Fig. 2.82 Using an axe to cut a wedge-shaped plug

Table 2.5 Stanley Surform blades

	Standard cut 21-505 Plane Planerfile Flat file	Fine cut 21-506 Plane Planerfile Flat file	Half round 21-507 Plane Planerfile Flat file	Metals & plastics 21-508 Plane Planerfile Flat file	Round 21-558 Round file	Fine cut 21-520 Block plane Ripping plane	Curved 21-515 Shaver tool
Hardwoods	■	■				■	■
Softwoods	■	■				■	■
End grain		■				■	■
Chipboard	■	■				■	■
Plywood	■	■				■	■
Blockboard	■	■				■	■
Vinyl	■		■				■
Rubber	■		■				■
Plaster	■		■		■		■
Thermalite	■		■		■		■
Chalk	■		■		■		■
Glass fibres	■		■		■		■
Brass		■		■			
Lead		■		■			
Aluminium		■		■			
Copper		■		■			
Mild steel				■			
Plastic laminates		■		■			
Plastic fillers		■		■			
Nylon	■			■			
Linoleum	■				■		
Ceramics	■						

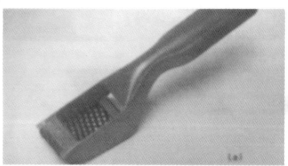

Fig. 2.83 Stanley Surform tools

2.8 Driving tools

These are tools which have been designed to apply a striking or turning force to fixing devices or cutting tools. For example;

a) hammers   – striking,
b) mallets   – striking,
c) screwdrivers   – turning,
d) carpenter's
    braces   – turning,
e) hand drills
    (wheel-
    braces)   – turning. } dealt with under *boring tools*

Hammers

There are four types of hammer you should become familiar with;

1   claw hammers,
2   Warrington or cross-pein hammers,
3   engineer's or ball-pein hammers,
4   club or lump hammers.

Claw hammers (Fig. 2.84) These are steel-headed with a shaft of wood or steel, and a handgrip of rubber or leather. They are used for driving medium to large nails and are capable of withdrawing them with the claw. Figure 2.85 shows a claw hammer being used to withdraw a nail – notice the waste wood used both to protect the workpiece and to increase leverage.

This type of hammer is the obvious choice for site workers involved with medium to heavy constructional work. It can, however, prove cumbersome as the size of nail decreases.

Warrington or cross-pein hammer
(Fig. 2.84) Although capable of driving large nails, this is better suited to the middle to lower range, where its cross pein enables nails to be started more easily. This hammer is noted for its ease of handling and good balance.

Figure 2.86 shows a Warrington hammer being used to demonstrate that, by using the full length of its shaft, less effort is required and greater accuracy is maintained between blows – thus increasing its efficiency. (This applies to all hammers and mallets.)

Tools associated with this hammer are the nail punch (featured in Fig. 11.4) and pincers. Pincers provide the means to withdraw the smallest of nails.

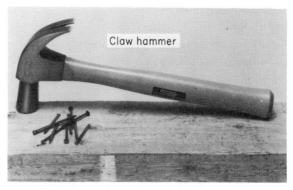

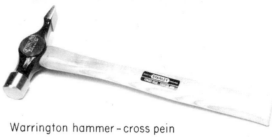

Warrington hammer – cross pein

Fig. 2.84 Hammers

Engineer's or ball-pein hammers
(Fig. 2.84) The larger sizes are useful as general-purpose heavy hammers and can be used in conjunction with wall-plugging chisels etc.

Club or lump hammer (Fig. 2.84) Used mainly by stone masons and bricklayers, it is, however, a useful addition to your tool kit as a heavy hammer capable of working in awkward and/or confined spaces.

Warning: Hammer heads should never be struck against one another or any hardened metal surface, as this action could result in the head either splitting or splintering – particles could damage your eyes.

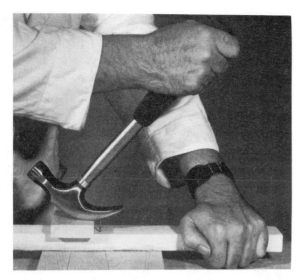

Fig. 2.85 Claw hammer withdrawing a nail. Leverage should be terminated and hammer repositioned by using a packing (as shown) before the hammer shaft reaches the upright position, or an angle of 90° to the face of the workpiece - otherwise undue stress could damage the hammer shaft

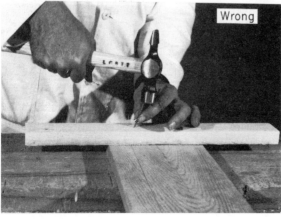

Fig. 2.86 Holding a hammer

If a hammer face becomes greasy or sticky, the chances are that your fingers or workpieces will suffer a glancing blow. Always keep the hammer's striking face clean by drawing it across a fine abrasive paper several times.

Wooden shafts are still preferred by many craftsmen, probably because of their light weight and good shock-absorbing qualities. Users of wooden-shafted hammers must however make periodic checks to ensure that the head is secure and that there are no hair-line fractures in the shaft. Figure 2.87 illustrates three examples of how a shaft could become damaged.

Mallets

The head of a mallet, which provides a large striking face, and its shaft, which is self-tightening (tapered from head to handle), are usually made from beechwood and weigh between 0.4 kg and 0.6 kg – choice will depend on the mallet's use and the user. Many joiners prefer to make their own mallet, in which case it can be made to suit their own hand.

The joiner's mallet should be used solely to strike the handle of wood-cutting chisels – Fig. 2.88 shows its correct use, and Fig. 2.81

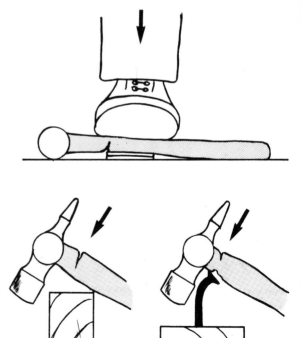

Fig. 2.87 Damage to a wooden shaft

shows it in use. Using a mallet for knocking together timber frames or joints should be regarded as bad practice, because – unless protection is offered to the surface being struck – the mallet will have a similar bruising effect to a hammer.

Screwdrivers

The type and size of screwdriver used should relate not only to the type and size of screw but also to the speed of application and the location and quality of the work.

There are three basic types of screwdriver used by the carpenter and joiner;

1 the fixed or rigid-blade screwdriver,
2 the ratchet screwdriver,
3 the spiral ratchet or pump screwdriver.

Each is capable of tackling most if not all of the screws described in Section 15.2.

Rigid-blade screwdrivers These are available in many different styles and blade lengths, with points to suit any screw head. They work directly on the screw head (screw eye) to give positive driving control. Figure 2.89 illustrates two types of rigid-blade screwdriver.

Ratchet screwdriver Available to handle slotted and superdriv (posidriv) headed screws and is operated by rotating its firmly gripped handle through 90° – then back – and repeating this action for the duration of the screw's drive. A clockwise or counter-clockwise motion will depend on the ratchet setting – a small sliding button, illustrated in Fig. 2.90, is used to pre-select any of the following three operations;

1 forward position   – clockwise motion,
2 central position   – rigid blade,
3 backward position – counter-clockwise
                           motion.

Because the driving hand always retains its grip on the screwdriver throughout its operation, this ratchet facility speeds up the process.

Spiral ratchet screwdriver This is often termed a *pump screwdriver*, because of its pump action, and is by far the quickest hand method of driving screws. Not only can it handle all types of screws, it can also be adapted to drill and countersink holes.

Figure 2.91 shows a Stanley *Yankee* spiral ratchet screwdriver, and Fig. 2.92 indicates some of the many accessories available. Its ratchet control mechanism is similar to that of the standard ratchet

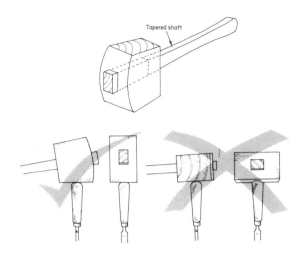

Fig. 2.88 Using a mallet correctly

Pozidriv screwdriver

Slotted screwdrivers

Fig. 2.89 Rigid-blade screwdrivers

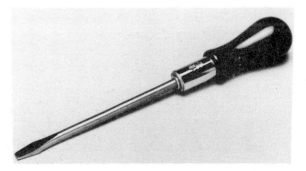

Fig. 2.90 Ratchet screwdriver

screwdriver. However, its driving action is produced by pushing (compressing) its spring-loaded barrel over a spiral drive shaft, thus rotating the chuck (bit-holder) every time this action is repeated.

Warning: by turning its knurled locking collar, the spiral drive shaft can be fully retained in the barrel, enabling it to be used as a short rigid or ratchet screwdriver. But, while the shaft is spring-loaded in this position, its point must always be directed away from the operator, as it is possible for the locking device to become disengaged, in which case

the shaft will lunge forward at an alarming rate and could result in serious damage or injury.

After use, always leave this screwdriver with its spiral shaft fully extended. The spring should never be left in compression.

Screwdriver efficiency   With the exception of the *stub* (short-blade) screwdriver, the length of blade will correspond to its point size. If driving is to be both effective and efficient, it is therefore important that the point (blade) must fit the screw eye correctly. Figure 2.93 illustrates how face contact with a slotted eye can affect the driving efficiency – (a) and (b) are inclined to come out of the slot; therefore (c) should be maintained at all times.

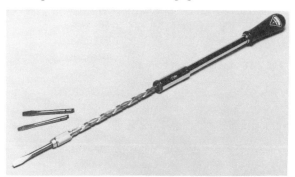

Fig. 2.91   Spiral ratchet screwdriver and bits

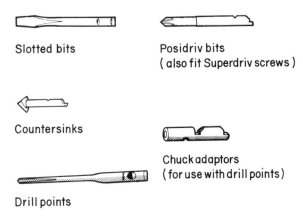

Slotted bits

Posidriv bits
(also fit Superdriv screws)

Countersinks

Chuck adaptors
(for use with drill points)

Drill points

Fig. 2.92   Some spiral ratchet screwdriver accessories

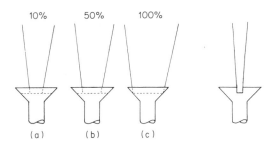

Fig. 2.93   A guide to slot screwdriver efficiency

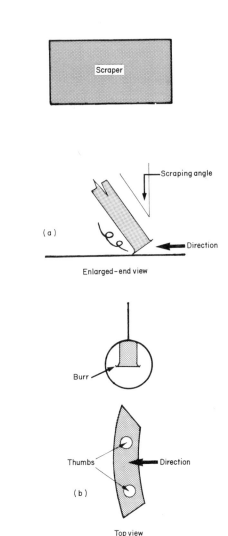

Fig. 2.94   Hand-held cabinet scraper

Table 2.6 Comparative grading of abrasive sheets

GRAIN	BACKING	PRODUCT CODE	BOND	COAT	SHEETS	COILS	ROLLS	BELTS	DISCS	1200	1000	800	600	500	400	360	320	280	240	220	180	150	120	100	80	60	50	40	36	30	24	16
GLASS	PAPER 'C'	117	G	C	●														FL	0		1	1½	F2		M2*		S2	2½	3		
	PAPER 'C'	121	G	C	●														00	0		1	1½	F2		M2*		S2	2½	3		
GARNET	PAPER 'A'	129	G	0	●											9/0		7/0	6/0	5/0	4/0	3/0	2/0	0							
	PAPER 'C' + 'D'	128	G	0	●																5/0	4/0	3/0	2/0	0	½	1	1½				
ALUMINIUM OXIDE	PAPER 'C'	134	G	0	●																●	●	●	●	●	●	●	●	●	●		
	PAPER 'E'	156	R/G	C	●		●	●	●						●		●	●	●	●	●	●	●	●	●	●	●	●	●	●	●	
	PAPER 'E'	157	R	C				●	●												●	●	●	●	●	●	●	●	●	●		
	CLOTH 'X'	141	R	0	●	●	●	●	●												●	●	●	●	●	●	●	●	●	●		
	CLOTH 'X'	147	R	C			●	●	●								●	●	●	●	●	●	●	●	●	●	●	●			●	
	CLOTH 'FJ'	144	R	C				●						●	●	●	●	●	●	●	●	●	●						
	CLOTH 'J'	145	R	C				●								●	●	●	●	●	●	●	●	●						

KEY

BOND

| G | GLUE | | R | RESIN | | R/G | RESIN OVER GLUE |

GRAIN CONFIGURATION

| O | OPEN COAT | | C | CLOSE COAT |

GRAIN SIZE
*IF = 350 GRIT M2 = 70 GRIT

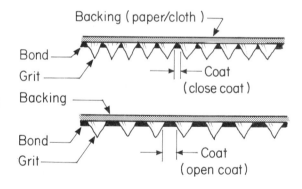

Backing (paper/cloth)
Bond
Grit
Coat (close coat)

Backing
Bond
Grit
Coat (open coat)

2.9 Finishing tools and abrasives

The final cleaning-up process will be determined by whether the grain of the wood is to receive a transparent protective coating or is to be obscured by paint. In the former case, treatment will depend on the type and species of wood; whereas the treatment for painting is common to most woods.

Hardwood Because its main use is decorative, hardwood is usually given a transparent protective coating. Unfortunately, the grain pattern of many hardwoods makes them difficult to work – planing often results in torn or ragged grain. A scraper can

ABRASIVES FOR HAND USE

GLASS PAPER 'C'	Cabinet quality for joinery and general wood preparation.	**117**
GLASS PAPER 'C'	Industrial quality for joinery and general wood preparation.	**121**
GARNET PAPER 'A'	For fine finishing and contour work on furniture.	**129**
GARNET PAPER 'C' + 'D'	Durable product for fine finish stages of wood preparation.	**128**

ABRASIVES FOR GENERAL MACHINE USE

ALUMINIUM OXIDE PAPER 'C'	For rapid stock removal with reduced clogging. For hand and orbital sander use.	**134**
ALUMINIUM OXIDE PAPER 'E'	For sanding medium and soft woods. A durable product recommended for heavy hand sanding, orbital sanders, and pad and wide belt sanders.	**156**
ALUMINIUM OXIDE CLOTH 'X'	General belt sanding of wood, particularly recommended for portable sanders, open coat configuration reduces clogging.	**141**

resolve this problem, and should be used before finally rubbing down the surface with abrasive paper. Note: always follow the direction of the grain when using abrasive paper before applying a transparent finish (see Fig. 2.96).

Softwood   Softwood surfaces that require protection are usually painted – there are, however, some exceptions. In preparing a surface for paint, flat surfaces must be flat but not necessarily smooth – unlike with transparent surface treatments, minor grain blemishes etc. will not show. Surface ripples caused by planing machines do show, however, and will require levelling with a smoothing plane, after which an

abrasive paper should be used. The small scratch marks left by the abrasive help to form a key between the wood and its priming paint (first sealing coat).

Scraper

This is a piece of hardened steel sheet, the edges of which have been turned to form a *burr* (cutting edge), see Fig. 2.94(a).

A scraper is either of a type which is held and worked by hand or is set into a scraper plane which looks and is held, like a large spokeshave. Flat scrapers must always be kept bent during use, to

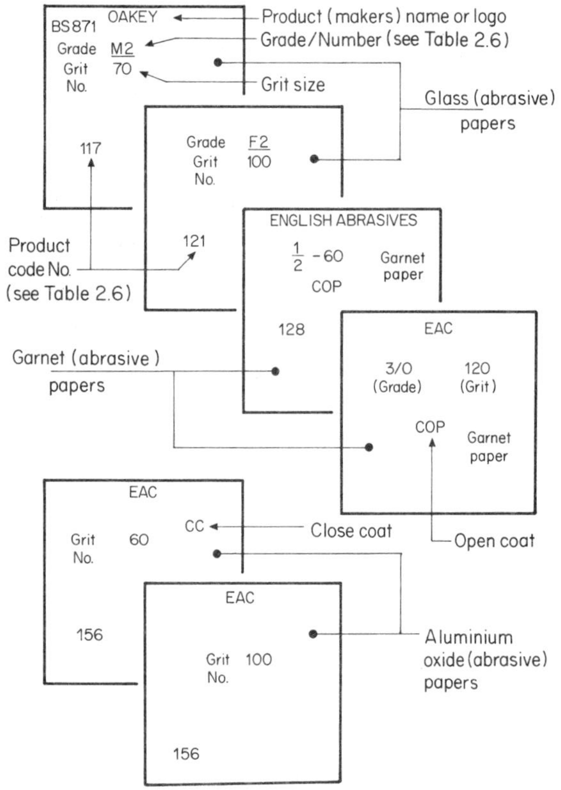

Fig. 2.95 Understanding the letters/numbers and coding used on the backing of coated abrasives as identification

Choosing the correct coated abrasive for the type of work

Key	Application
Backing	
Paper	A lower cost material for less demanding uses.
Cloth	Higher tensile strength, strong and durable.
Bond	
(G) All Glue	Used for hand and light machine applications, a flexible bond.
(R/G) Resin over Glue	Intermediate flexibility and durability.
(R) All Resin	Higher band strength and heat resistance, for more demanding applications.
Grain Configuration	
(O) Open coat	Reduced clogging, coarser finish.
(C) Closed coat	Finer finish, faster cutting.

Grit size

The coarser the grit (larger particles), the smaller the grit size number, according to international FEPA 'P' standards. The non-technical abrasives are graded according to different standards and are equivalent as shown in table.

Note: grit (particle) size :– the smaller the grit number the more coarse the grit. For example:

Grit No. 24   very coarse
Grit No. 50 – 70 medium
Grit No. 400   very fine

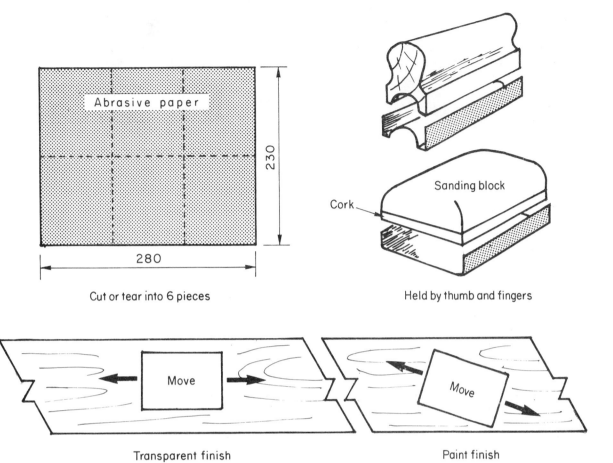

Cut or tear into 6 pieces

Held by thumb and fingers

Transparent finish

Paint finish

Fig. 2.96 Sanding by hand

avoid digging their sharp corners into the wood
(Fig. 2.94(b)). Scraper planes, however, can be
pre-set to the required cutting angle and simply
require pushing.

Sanding

Sanding is the application of abrasive-coated paper
or cloth to the surface of wood.

There are several kinds of grit used in the
manufacture of abrasive sheets, the two most
popular being glass and garnet. Both are available
in a sheet size 280 mm × 230 mm and are graded
according to their grit size, which ultimately
determines the smoothness of the wood, see Table
2.6 and Fig. 2.95.

A sanding block should always be used to ensure
a uniform surface, be it flat or round. Typical
examples are shown in Fig. 2.96, together with a
method of dividing a sanding sheet into the
appropriate number of pieces to suit the blocks.

Table 2.7 Holding equipment

Equipment	Use	
	Preparing material	Assembly aid
Holding tool		
a) Bench vice	Yes	Some models
b) Bench holdfast (Fig. 2.97)	Yes	No
c) G cramp (Fig. 2.98)	Yes	Yes
d) L cramp (Fig. 2.101)	Yes	Yes
e) Sash cramp (Fig. 2.99)	Yes	Yes
f) Dowelling jig	Yes	
g) T bar cramp (Fig. 2.100)	No	Yes
Holding devices (Fig. 2.102)		
h) Bench stop*	Yes	No
i) Bench hook*	Yes	No
j) Mitre block	Yes	No
k) Mitre box	Yes	No
l) Dowel cradle	Yes	No
m) Saw stool (trestle)	Yes	Yes

* Provision can be made in their construction for left- or right-handed users.

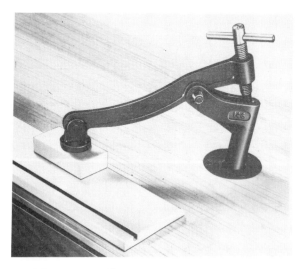

Fig. 2.97   Bench holdfast

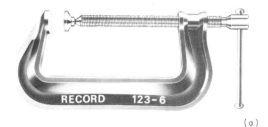

(a)

Fig. 2.98(a)   G cramp

(b)

Fig. 2.98(b)   G cramp and bench holdfast

Fig. 2.99   Sash cramp and lengthening bar

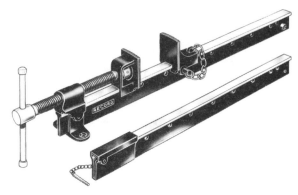

Fig. 2.100   T bar cramp and lengthening bar

2.10 Holding equipment (tools and devices)

Holding tools are general-purpose mechanical aids used in the preparation and assembly of timber components. Holding devices have usually been contrived to meet the needs of a specific job or process, and Table 2.7 gives examples.

Items (a) to (i) in Table 2.7 are mentioned throughout the text and are almost self-explanatory. The mitre block and box, (j) and (k), are devices used to support squared or moulded sections while they are sawn to an angle of 45° degrees. The dowel cradle, (l), holds squared or rounded sections. Saw stools or trestles, (m), are usually used in pairs, either to form a low bench or for support while sawing.

Figure 2.101 features a double-sided work bench, together with items (a), (h), (i), (j), (k), (l) and (m).

Fig. 2.101   L Cramp

2.11 Tool storage and containers

The variety and condition of the tools used by the craftsperson often reflects the quality and type of work capable of being under taken by them. Apart from your basic tool kit of, say,

a) four-fold metre rule,
b) flexible steel tape,

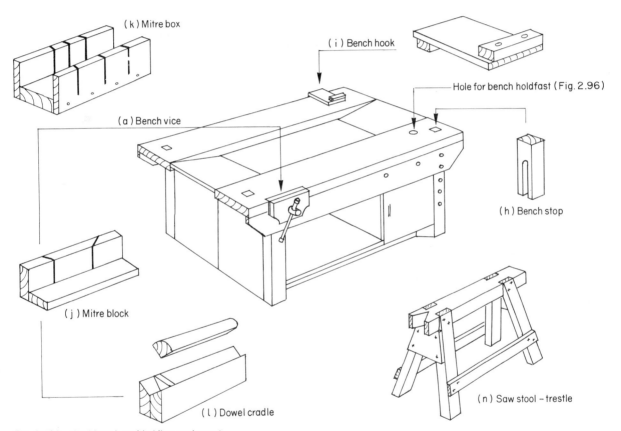

Fig. 2.102 Workbench and holding equipment

c) combination square,
d) claw hammer,
e) panel saw, and possibly a Tenon saw,
f) jack and smoothing planes,
g) assortment of chisels firmer/bevel-edged,
h) carpenter's brace,
i) assortment of twistbits and countersink,
j) hand drill (wheelbrace),
k) assortment of twistdrills,
l) screwdrivers,
m) bradawl.

(Note: sharpening accessories (oilstones etc.) should also be included on this kit.)

There is no doubt that irrespective of the kind of work you are employed to do, over the forthcoming years many different types of tool will be acquired – all of which will require some form of protection and safe storage provision – whether you are to be site- or workshop-based.

Enlarging the immediate number and type of tools in your kit will depend on your job specification. Broadly speaking, tool kits can be broken down into three groups (see Tables 2.1–5):

1 *Everyday use* – basic tool kit.
2 *Occasional use* – more common of the special use tools.
3 *Specific use* – special use tools.

In time you will find that you are duplicating some of your everyday tool kit. For example; having two panel saws should ensure that one is always kept sharp.

No matter what means of storage is chosen, the most important factors to consider about a container are that each item is housed separately or is individually protected from being knocked against another, and that all cutting edges are withdrawn or sheathed in some way. Of course the body of the tool must also be protected from damage – unfortunately many joiners seem to

regard this form of protection as secondary. Always bear in mind that a well worn tool is not one sporting the battle scars of its container but which, after many years of active service arrives at that point in time virtually unscathed.

There are a variety of ways by which you may want to house your tools – the method usually chosen often reflects the type of work your company undertakes. The final choice is usually yours. Different methods of containing tools include;

a) traditional toolbox (Fig. 2.103(a))
b) traditional toolchest (Fig. 2.103(b))
c) tool case (Fig. 2.103(c))
d) tool tray (usually racked) (Fig. 2.103(a))
e) tool bag (Fig. 2.103(e))
f) joiner's bass (Fig. 2.103(f))

A combination of the above is usually chosen.

Fig. 2.103 Different methods of containing tools

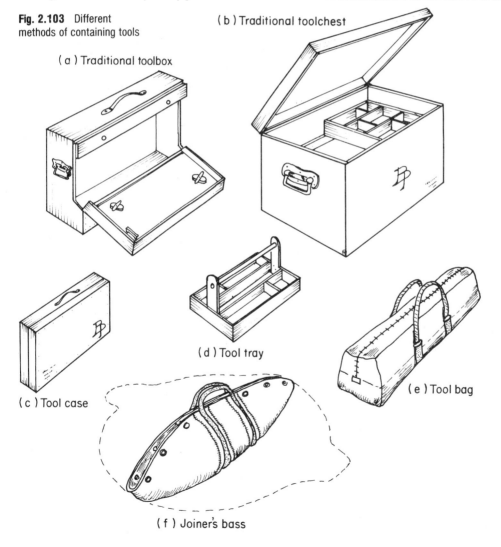

(a) Traditional toolbox

(b) Traditional toolchest

(c) Tool case

(d) Tool tray

(e) Tool bag

(f) Joiner's bass

The traditional toolbox (Fig. 2.103(a)) – a strong rectangular drop fronted box long enough to house the longest handsaw – generally suitable for both workshop and site use. Top, bottom and sides of timber are dovetail jointed at the corners. Plywood is glued, screwed and/or nailed to this framework to form the front and back. Once set, a front portion is cut out to form a hinged flap – this flap houses the handsaws. The interior can accommodate one or more drawers.

I have always regarded this arrangement as unsatisfactory for a number of reasons, the two main ones being:

1 In order to gain access to the tools the flap has to take up a lot of floor space, and if left open is a tripping hazard to passers by.
2 Because the top of the box is often used as a saw stool, and seat during break periods, the handle and closing/security mechanism is in the way.

It was with these factors in mind that I designed and developed a system of boxes known as the *Porterbox System* which collectively can be found in Chapter 3.

The traditional toolchest (Fig. 2.103(b)) – a very strong, secure, top opening rectangular box of solid construction. It is designed to accommodate all of the joiners tools (everyday and special usage). Larger tools are housed within base compartments, whereas smaller tools are separated within a series of compartmentalized sliding lift-out trays. The two upper trays being half the box length, thereby allowing access to box contents without their removal.

Transporting such a chest is a two-man operation, therefore heavy duty drop-down chest handles bolted one to each end are essential.

Tool case (Fig. 2.103(c))   There are occasions when the intended job does not warrant taking all your basic toolkit. On these occasions a smaller, lighter box (scaled down version of your traditional tool box) may be the answer. In this case the length of the saw may not be the main criterion in its design, because a separate saw case (see fig, 2.104a) could be used along with the tool case. However a tenon saw should be capable of being fitted within the lid of the tool case.

Tool tray (Fig. 2.103(d)) – an open top box with a box length carrying handle, used as a means of transporting your tools from job to job within the confines of the work area (site). It may also serve

as a means of keeping tools tidy when working with other trades, a rack for chisels etc. is a useful feature. After each work period tools are then transferred back to their secure main tool box, or chest.

Tool bag (Fig. 2.103(e)) – an elongated bag of hard wearing material long enough to contain a hand saw, and strong enough to hold a variety of tools. Its great advantage over a box or case is its lightness and transportability. Its disadvantage must be that each tool contained therein needs individual protection and that their level of security is at a minimum.

Traditional joiners bass (Fig. 2.103(f)) – a near circular shape of heavy duty canvas type fabric with two roped handles attached across the fabric which, when brought together, form the shape of a half-moon bag – leather straps, or eyelets around the edge lashed with cord provide the means of closing and securing the bass. The only advantage over the bag, I feel, is that when the bass is fully open all its contents, with the exception of those contained in side pockets, pouches, or rapped in rolls etc., are instantly revealed.

Note: the bass used by a plumber is usually smaller in diameter.

Individual tool storage

All tools should be restricted from any movement when being stored, particularly if they are to be

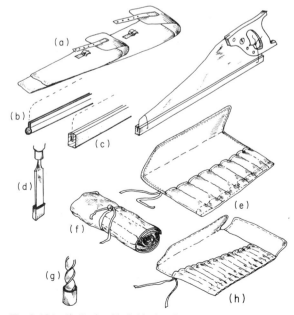

Fig. 2.104 Methods of individual tool storage

transported. (The cutting edges and sharp points are the most vulnerable to damage.) All planes should have the cutting face of their blades retracted before storage. Other tools requiring special treatment are;

1   saws (saw teeth),
2   chisels (cutting edge),
3   twist bits (cutters, spurs(wings), and thread).

Saws   Where box protection is not available, a saw bag (Fig. 2.104(a)) made from a strong material is a good alternative. In both cases the teeth will still need protecting – protection can be provided by plastics sheath (Fig. 2.104(b)) available from the saw manufacturer and tool suppliers, alternatively a timber lath with a groove down the length of one edge (Fig 2.104(c) and held onto the saw by a band will do just as well; or perhaps you could design your own.

Chisels   Most new chisels are now provided with a plastics end cover(sheath) (Fig. 2.104(d)) by their manufacturer to safely protect the cutting edge of each blade – end covers are now available (usually in sets) from suppliers as a separate item.

Where chisels are not racked or safely compartmentalised, a strong purpose made chisel roll is a useful storage alternative. As shown in Fig. 2.104(e), when unrolled selection is quick and easy as each chisel is easily identified within its own individual pocket.

On packing away, simply fold down the top flap over the handles, roll up (Fig. 2.104(f)), and secure the roll by using the ties or staps provided.

Note: all chisels must have their cutting edges safely covered with an end protection before being put into a chisel roll.

Twist bits   End protection is essential – plastic sheathes (Fig. 2.104(g)) are available. Unlike chisels, the cutting edges of the cutters only have a limited resharpening capability. If the spurs

(wings) or thread become damaged the cutting effectiveness is reduced.

Bit rolls (Fig. 2.104(h)) are purpose-made bit holders. Similar in design and function to chisel rolls, the main difference is that the pockets are made slimmer to accommodate individual bits.

2.12   Tool maintenance

Of all the cutting tools mentioned, only saws and Surforms are purchased ready for use – the remainder will need sharpening.

Once tools have been sharpened, it is only a question of time before they become *dull* (blunt) again. Some makes of tools dull much quicker than others, due to the quality of steel used to make their blade or cutter, but general dulling is caused either by the type of work or by the abrasive nature of the material being cut. However, other contributory factors include foreign bodies encased in the material being cut, for example hidden nails or screws.

Keeping tools sharp will require varying amounts of skill in the use of those tools and devices associated with tool maintenance. The techniques used for tool maintenance can differ from craftsman to craftsman, and in some cases take many years to perfect. Table 2.8 gives a list of equipment thought necessary for the maintenance of the cutting tools previously discussed.

Equipment

Grinding machines   Every time a blade is sharpened, part of its grinding angle (Fig. 2.114) is worn away, making sharpening more difficult. The grinding angle must therefore be re-formed by machine. The grinding machine, together with the relevant regulations, is dealt with in Chapter 5.

Oilstones   These are used to produce a fine cutting edge (a process called *honing*). They are

Table 2.8   Tool-maintenance equipment

Equipment	Tool					
	Drills	Saws	Planes	Bits	Chisels	Screwdrivers
Grindstone (machine)	Yes	-	Yes	-	Yes	Yes
Oilstone	-	-	Yes	-	Yes	Yes
Slipstone	-	Yes	Yes	-	Yes	-
Stropstick	-	-	Yes	-	Yes	-
Oil can	-	-	Yes	-	Yes	-
Mill saw file	-	Yes	-	-	-	-
Saw file	-	Yes	-	Yes	-	-
Needle file	-	-	-	Yes	-	-
Saw vice-clamp	-	Yes	-	-	-	-

generally manufactured from one of three abrasive-grit sizes, producing stones of either fine, medium, or coarse texture. It is possible to purchase a stone with a coarse surface on one side and a fine on the other, called a *combination stone*.

Oilstones are very brittle and will break or chip unless they are protected. Figure 2.105 illustrates a method of constructing a protective box out of hardwood – half of the oilstone is mortised into the base and half into the lid, and allowance *can* be made for packings at the ends to enable the full length of the stone to be used (see Fig. 2.116). Rubber washers or blocks have been let into or pinned on to the underside to prevent the box sliding about during use.

Slipstones These are composed in a similar manner to oilstones, but are designed to hone the curved cutting edges of gouges and moulding-plane cutters etc. Figure 2.121 shows a slipstone in use – hand held. If a holding device were used, it would leave both hands free to position the gouge or cutter against the stone. Such a device is illustrated in Fig. 2.106 and can also serve as a protective box.

Stropstick This gives a blade its final edge. It is made from a thick short length of leather strap, about 50 mm wide, stuck or fixed to a flat board (see Fig. 2.116).

Oil Oil provides the stone with the necessary lubrication and prevents the pores of the stone becoming clogged with particles of grit or metal. Only good-quality machine oil should be used; but the very occasional use of paraffin as a substitute often re-establishes the stone's keenness to cut.

Files These are required to sharpen saws and bits. At least two sizes of saw files will be required, together with a flat or mill saw file. Saw files can also be used to sharpen twist bits, but one or two small needle files would prove useful for the smaller sizes of bit. Figure 2.107 shows the different file sections.

Holding equipment There are various aids for holding a tool or cutter while it is being sharpened. When sharpening a saw, it is important that it is held firm and positioned in such a way that the teeth can be seen – preferably without stooping. Figure 2.108 illustrates a device which is ideally suitable for this purpose when either fixed into a bench vice or cramped to a solid object.

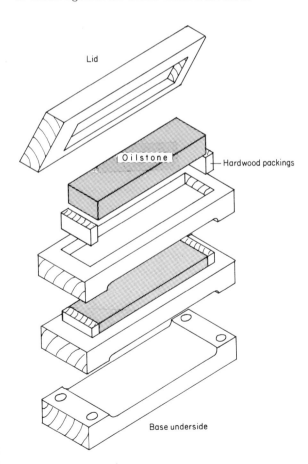

Fig. 2.105 Oilstone box

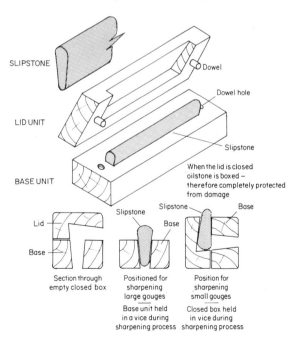

Fig. 2.106 Slipstone box and holding device

Sharpening a spokeshave blade can be difficult and dangerous, due to its shortness. A holder, as shown in Fig. 2.109, provides the answer.

Saws

Of all tools, saws are the most difficult to sharpen. The problem can be lessened if sharpening is carried out at frequent intervals, because as the condition of the saw worsens, so does the task of resharpening. Signs of dullness (bluntness) are the teeth tips becoming shiny, or extra pressure being needed during sawing. The sharpening technique will vary with the type of saw.

If teeth become very distorted, due to inaccurate sharpening or accidentally sawing nails etc., then the following processes should be undertaken;

1 topping,
2 shaping,
3 setting,
4 sharpening.

Topping – bringing all the teeth points in line (Fig. 2.110(a)) by lightly drawing a long mill saw file over them. A suitable holding device is shown in Fig. 2.110(b).

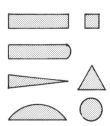

Fig. 2.107 File sections

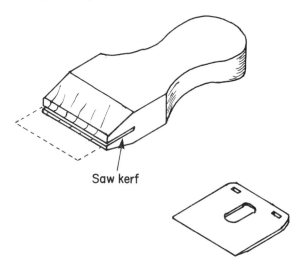

Saw kerf

Fig. 2.109 Spokeshave blade holder

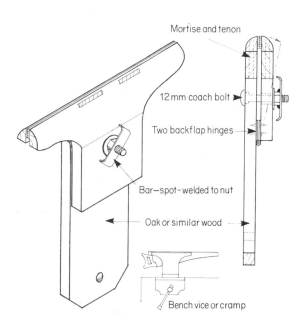

Mortise and tenon

12 mm coach bolt

Two backflap hinges

Bar–spot–welded to nut

Oak or similar wood

Bench vice or cramp

Fig. 2.108 Saw vice

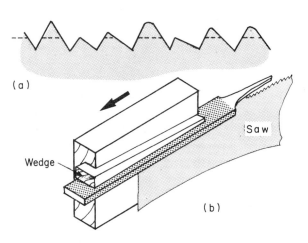

(a)

Saw

Wedge

(b)

Fig. 2.110 Topping a saw

Shaping – restoring the teeth to their original shape and size. The file face should be just more than twice the depth of the teeth and held level and square to the saw blade throughout the process (see Fig. 2.111). Shaping is the most difficult process to perfect.

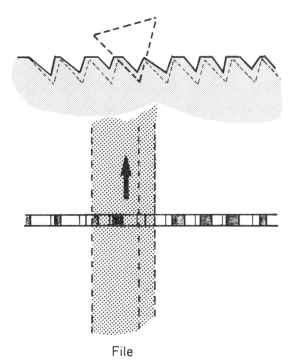

File

Fig. 2.111 Shaping saw teeth

Fig. 2.112 Setting saw teeth. Note: teeth should be covered with a saw blade sheath (Fig 2.104(b) or (c))

Setting – bending over the tips of the teeth to give the saw blade clearance in its kerf (see Fig. 2.23). Manufacturers and *saw doctors* use a cross-pein hammer and a saw anvil for this purpose. A suitable alternative is to use saw set pliers, which can be operated with one hand simply by pre-selecting the required points per 25 mm of blade, placing over each alternate tooth, and squeezing (Fig. 2.112).

Sharpening – putting the cutting edge on the teeth. Cramp the saw, with its handle to your right, as low as practicable in its vice. Saw teeth of the over hanging part of the saw should be covered (use saw blade sheath – Fig. 2.104(b)). Place your file against the front face of the first 'toe' tooth set towards you and the back edge of the tooth facing away from you, then angle the file to suit the type of saw or work (Fig. 2.113(a)) and, keeping it flat (Fig. 2.113(b)) and working from left to right, lightly file each alternate V two or three times (Fig. 2.113(c) and (d)). After reaching the handle (heel), turn the saw through 180° and repeat as before, only this time working from right to left (Fig. 2.113(c) and (e)).

If the teeth are in a very bad condition, you would be well advised to send your saw to a saw doctor (a specialist in saw maintenance) until you have mastered all the above processes.

Note: whenever a saw is not in use, ensure that its teeth are protected.

Planes

Plane *irons* (blades) require a grinding angle of 25° and a *honing* (sharpening) angle of 30° (Fig. 2.114) if they are to function efficiently – with one exception: Plough planes use a common 35° grinding and honing angle.

The blade shape should correspond to one of those shown in Fig. 2.115. A smoothing-plane blade has its corners rounded to prevent digging in, whereas a jack-plane blade is slightly rounded to encourage quick easy removal of wood. A rebate-plane blade must be square to the iron, for obvious reasons.

Honing Honing (sharpening) using an oilstone requires much care and attention. The following stages should be carefully studied.

1 Remove the cap iron from the cutting iron – use a large screwdriver to remove the cap screw, *not* the lever cap, otherwise its chrome coating will soon start to peel.

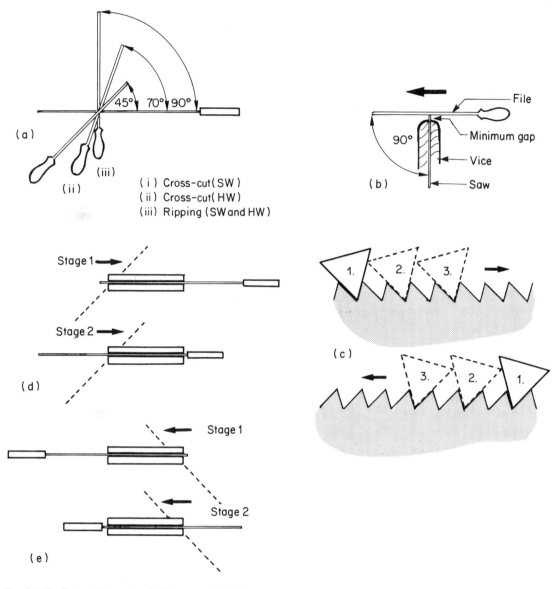

(a)

45° 70° 90°

(iii)
(ii)

(i) Cross-cut (SW)
(ii) Cross-cut (HW)
(iii) Ripping (SW and HW)

(b)

File
90°
Minimum gap
Vice
Saw

Stage 1 →
Stage 2 →

(d)

(c)
1. 2. 3.
3. 2. 1.

Stage 1 ←
Stage 2 ←

(e)

Fig. 2.113 Sharpening saw teeth. Note: saw teeth of the overhanging portion of the saw should be covered with a saw blade sheath to protect the operator from the sharp teeth

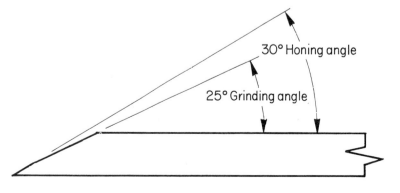

30° Honing angle
25° Grinding angle

Fig. 2.114 Plane blade - grinding and honing angles

Note: this operation should only be carried out whilst the cap iron and cutting iron are laid flat on the bench with some provision made to prevent the cutting edge rotating dangerously – for example, held within and up against the side of the bench wall.

2 Ensure that the oilstone is firmly held or supported.

3 Apply a small amount of oil to the surface of the oilstone.

4 Holding the cutting iron firmly in both hands, position its grinding angle flat on the stone, then lift it slightly – aim at 5° (Fig. 2.116(a)).

5 Slowly move the blade forwards and backwards (Fig. 2.116(b)) until a small burr has formed at the back. It is important that all the oilstone's surface is covered by this action, to avoid hollowing it (Fig. 2.116(c)).

6 The burr is removed by holding the iron *perfectly flat*, then pushing its blade over the oilstone two or three times (Fig. 2.116(d)).

7 The wire edge left by the burr can be removed by pushing the blade across the corner of a piece of waste wood (Fig. 2.116(e)).

8 If a white line (dullness) is visible on the sharpened edge, repeat stages 5–7. If not, proceed to stage 9.

9 Holding the iron as if it were on the oilstone, draw it over the strop stick (Fig. 2.116(f)).

The practice of showing off by using the palm of the hand as a strop is dangerous and silly – it can not only result in a cut hand or wrist, but may also cause metal splinters to

enter the skin, not to mention the associated problem of oil on the skin.

Blade positions   Bench-plane irons should be

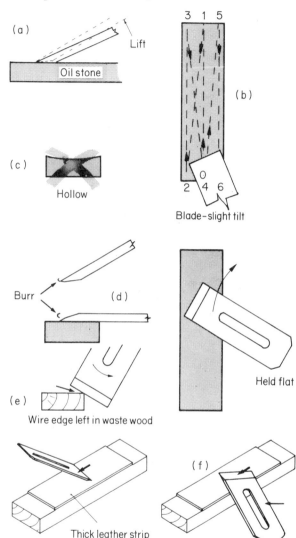

Fig. 2.116 Plane blade sharpening process

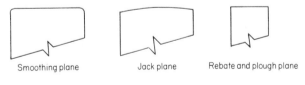

Smoothing plane   Jack plane   Rebate and plough plane

Fig. 2.115 Plane blade shapes

Fig. 2.117 Plane back irons and mouth opening adjustments

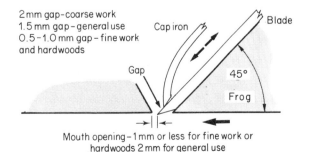

positioned as illustrated in Fig. 2.117 to ensure effective cutting. The position and angles of cutting irons without cap irons can vary, as can be seen in Fig. 2.118, which shows the arrangement for rebate, plough, and block planes.

Chisels

Flat-faced chisels   These should be ground and honed to suit the wood they are to cut, as shown in Fig. 2.119. However, where access to a grinding machine is difficult, i.e. on-site work, it is often possible to extend the useful life of the grinding angle as shown in Fig. 2.120.

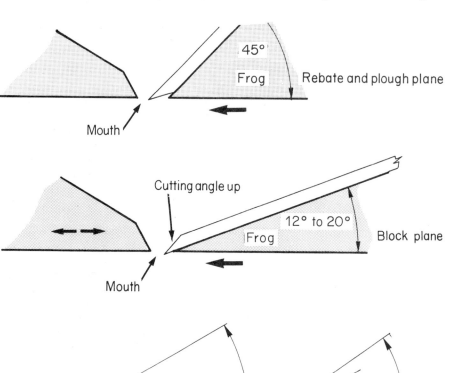

Fig. 2.118 Rebate, plough and block plane blade arrangement

Grinding and sharpening angles for softwood

Grinding and sharpening angles for hardwood

Fig. 2.119 Chisel grinding and honing angles

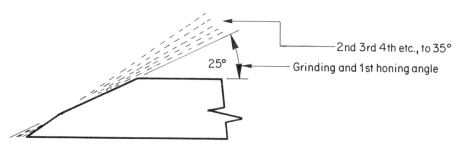

Fig. 2.120 Extending grinding angle life

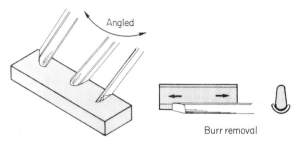

Fig. 2.121 Sharpening an out-cannel firmer gouge

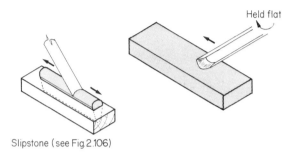

Fig. 2.122 Sharpening an in-cannel scribing gouge

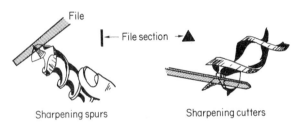

Fig. 2.123 Sharpening a twist bit

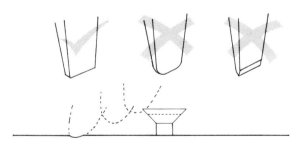

Fig. 2.124 Screwdrivers for slotted screws

The same principles of honing apply as for plane irons, although extra care must be taken not to hollow the oilstone.

Gouges Firmer gouges are ground on a conventional grinding machine – slowly rocking the blade over its abrasive surface. Figure 2.121 shows the method of honing and burr removal. Scribing gouges are ground on a special shaped grinder and honed on a slip stone as shown in Fig. 2.122.

Bits

A twist bit's life is considerably shortened every time it is sharpened, so be sure that sharpening is necessary. The spurs (wings) are usually the first to become dull, followed by the cutters. If the screw point becomes damaged, replace the bit. Figure 2.123 shows the method of sharpening a twist bit.

Screwdrivers

These are the most misused of all hand tools. They have been known to be used as levers, chisels, and scrapers. This is not only bad practice; it is also dangerous. Points do, however, become misshapen after long service, even with correct use. Figure 2.124 illustrates how a point can become misshapen, and possible consequences. The point should be re-formed to suit the screw eye by filing (using a fine-cut file), rubbing over an oilstone, or regrinding (see Section 5.9).

3

The Porterbox storage system

(First published in *Woodworker* magazine)

The Porterbox system as shown in Fig. 3.3 is a series of containers specifically designed to meet the storage needs of the carpenter and joiner, whether the requirement is for storing hand tools, power tools, accessories or hardware (ironmongery).

The system can be simply adapted to suit individual needs; for example the type of tools and equipment and the place of work, be it a workshop or on a building site.

The main component of the system is the *Porterbox* drop-fronted tool box. It has three optional add-on attachments:

1 *Portercaddy.*
2 *Portercase.*
3 *Porterdolly.*

These are then backed up by the *Porterchest* – originally designed to accommodate items of hardware, but just as at home holding an assortment of hand tools, or even portable powered hand tools – the choice is yours.

Practical project 1: Porterbox drop-fronted tool box and saw stool

This design is a practical alternative to the traditional joiner's tool box previously mentioned. The main idea behind this new concept was to allow all the established practices, such as leaving the front flap open for tool access during the work period and using the box top as a work surface and resting place during break periods etc., to continue, but in a safer and more practical way. This meant that:

1 The front flap would not project more than is necessary beyond the base of the box whilst still allowing easy access to the tools inside, thereby reducing the area of floor space taken up by the flap and the risk of passers-by tripping over it.

2 The box top would be capable of being used as an unobstructed flat work surface without the interference of a handle or latching mechanism.
3 Easy and quick access to racked tools would be available.

Other factors built into the design also meant that:

4 Construction would be strong without being too heavy.
5 It would be large enough to accommodate the basic tool kit together with a selection of special tools.
6 A single means of securely locking both top and side access in one operation would be provided.
7 The box would be tall enough and stable enough to use as a saw stool.
8 Its method of construction would be suitable for a first year trainee after having satisfactorily acquired the necessary proficiency in basic tool skills.
9 Both its interior (racks and shelves) and exterior (length and shape of top and overall box length etc.) could be easily adapted to suit individual needs.

Figure 3.1 shows details of a scaled-down of the tool box and its construction. A full cutting list is given in Table 3.1 in which all the various components are named and numbered to coincide with the drawing.

Method and sequence of construction

Rod Onto a large sheet of drawing paper, or piece of lining or wallpaper, attached onto a sheet of board material (for example, plywood or hardboard) with adhesive tape, draw at least two full-size sections through the box. For example:

a) a vertical lengthwise section (sectional front elevation) will, apart from showing constructional details ensure that the flap is long enough to receive your saw – adjust as necessary. *Do not forget to modify the cutting list accordingly.*

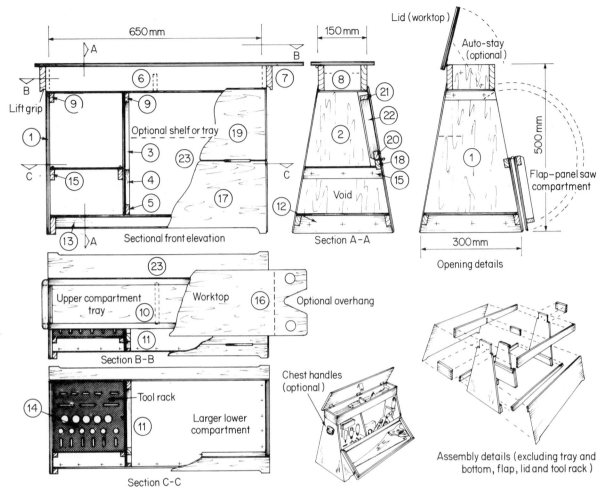

Fig. 3.1 Working drawing of Porterbox

Note:
1 Joints - all joints glued (PVA adhesive) and nailed or screwed (as stated). PLY/Hdbd back to carcase; glued, pinned and screwed at nlt 75 mm cts
2 Corners - all external corners to be bullnosed

3 Hinges - any butt type may be used
4 Lid - if used as a cutting board, hinged side should face the operative (to avoid tipping upwards)
5 Optional shelf - consider a sliding shelf by using plastics channel or drawer runner

b) a vertical section through A–A, which can also be used as a pattern to mark-off the shape of the plywood ends (1), and measure the width of the flap to the saw compartment.

Note: before setting-out any rod, always check the sectional size of the available material, and, providing the tolerance is acceptable these are the sizes which should be drawn onto the rod – *not those stated on the cutting list*.

Once the rod is complete in every detail (double check all the sizes, but these need not be shown on the rod), all future measurements, ie. lengths shapes etc, are taken by laying the material over the rod and then, with a sharp pencil, transferring

any length marks etc, onto the material. This process is known as *marking-off* or *marking-out*.

Not only does the method of using a rod save time, it dramatically cuts down the risk of mistakes, and thereby saves on materials.

Ends (1) (12 mm Plywood) Cut to shape, then while temporarily tacked together, plane to size and cut notches for tray sides. Then separate.

Partition (2) (10 mm Plywood) Using one end (1) as an outline template, cut to shape, not forgetting to allow for the front flap (19) and its clearance, then plane to size.

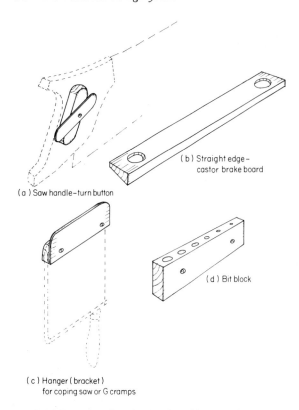

(b) Straight edge – castor brake board

(a) Saw handle – turn button

(d) Bit block

(c) Hanger (bracket) for coping saw or G cramps

Fig. 3.2 Examples of toolbox interior additions and attachments

Bearers (9) (12)(15)   Prepare, not forgetting to notch (12) to receive bearer (13). Fix to ends (1) with glue and screws.

Cleats (3)(4)(5)   Prepare, then glue and screw to partition (2).

Bearers (13)   To prepare, plane a bevel on the top edge to receive box bottom. Glue and screw to end of bearer (12).

Tray bottom (10) (6 mm Plywood)   Cut to size, fit, then glue and pin to bearers (9).

Box bottom (11) (6 mm Plywood)   Cut to size, fit, then glue and screw ($\frac{3}{4}$″ × 6 Csk) to bearers (12).

Front facing (17) (6 mm Plywood)   Cut to size, glue and screw ($\frac{3}{4}$″ × 6 Csk) to edging piece (18). Pin and glue facing to one end (1) (check that the carcase is square) then continue to pin and screw to glued faces of (1) and (13).
   Note: see also *mobility*.

Partition (2)   Fit around edging (18). Glue and nail (25 mm rounds) down though tray bottom (10), and up through box bottom (11).

Tray   Choose a suitable corner joint for members (6) and (7), for example:

a)   dovetail joint (double preferred),
b)   combed joint.

   The mid-division is optional; if a full-size spiral ratchet screwdriver is to be stored therein, leave out the division. Before assembly do not forget to cove one edge of each end member(7), to form the finger grip. When the corner joints are made and set, dress all the joints and clean-up the faces, then bull-nose the corners.
   Note: the tray could have been made before assembling the main carcase, but it was left until this stage, so that any small length variation could be taken into account.
   Slot the framework over ends (1). Glue the faces and receiving edges, then screw (1″ × 8 Csk) through plywood ends (1) into the handle ends (7), and up through the tray bottom ($\frac{3}{4}$″ × 8 Csk) into sides.

Back (23) (4 mm Plywood)   Cut to size, bevel top edge to fit tray side (6). Apply glue around all edges, pin/screw ($\frac{3}{4}$″ × 6 Csk) one side in position, adjust carcase for square, fix remaining sides.
   Note: see also *mobility*.

Carcase   Dress all corners flush, then bull-nose by sanding. Cut back and front toe piece to shape and sand. Note: if the box is to be used with the *Portercaddy/case* attachment, then the base of the box can be left flat (without the toepiece cut-out).

Flap (Drop-front) **(19) (6 mm Plywood)**   Fit over opening with a bevel to fit tray side (6). Mark position of ends (1), and bearers (9) (which must be cut back at an angle to allow for arc sweep of the flap stiffener).

Flap edging (20) and stiffeners (21 & 22) construct the framework. Use any suitable corner joint, for example:

Stiffener corners   –   Mitre with dovetailed key,
Stiffener to edging –   Bridle joint.

   Notch one end-stiffener to house point of panel saw. Apply glue around all stiffener edges, pin/screw ($\frac{3}{4}$″ × 6 Csk) plywood flap (19) onto the framework. Make turn-button keep (Fig. 3.2(a)), to suit saw handle – attach to flap with glue pins/screws.

Table 3.1 Cutting list for Porterbox

Drawing no.	Item description	No. off	Material	L	W	T	Remarks
1	End	2	Plywood	500	300	12	Multi-ply tapered
2	Partition	1	Plywood	370	280	9	Multi flap cut out
3	Partition cleat	1	SW	370	22	12	
4	Partition cleat	1	SW		22	12	
5	Partition cleat	1	SW	250	34	12	
6	Tray front and back	2	SW	715	70	18	
7	Tray end	2	SW	150	70	18	
8	Tray divider	1	PLY/SW	134	50	12	Optional
9	Tray bearers	3	SW	165	34	12	
10	Tray bottom	1	PLY	650	150	4-6	
11	Box bottom	1	PLY	650	280	6	
12	Box bearers	2	SW	300	*46	12	
13	Box bearers	2	SW	650	34	18	
14	Tool rack	1	PLY	220	220	4-6	
15	Tool rack bearers	2	SW	240	22	12	
16(a)	Lid (top)	1	PLY/MDF	800	190	12/18	Overhang - Optional
16(b)	Lid	1	SW	800	190	22	Overhang - Jointed
17	Front facing	1	PLY	676	220	6	
18	Front edging	1	SW	650	22	12	Hinge support
19	Flap (front)	1	PLY	676	220	6	
20	Flap edging	1	SW	650	22	12	Hinge support
21	Flap stiffener	1	HW	650	30	12	
22	Flap stiffener	2	HW	220	30	12	
23	Back	1	PLY/Hdbd	676	450	4-6	
Hardware							
24	Butt hinges	3	Steel	50			Uncranked (lid)
25	Butt hinges	3	Steel/Brass	50			Alternatives
26	Back flaps	3	Steel	50			Flap hinges
27	Hasp and staple	1	Steel	150			Use with pad lock
28	Chest handles	2	Steel	85			Optional

Key SW - Softwood eg. redwood, whitewood or similar species (± 1 mm) HW - Hardwood; straight grained species (± 1 mm) PLY - Plywood; exterior quality not necessary MDF - Medium density fibre-board Hdbd - Hardboard (good quality); less expensive substitute for plywood * - Use 34 × 12 mm if base units *Portercaddy* or *Portercase* are to be added.

Hinging flap

Semi-concealed: use one and a half pairs of 50 mm steel/brass butt hinges (25), or hurl-hinges if recessing is to be avoided.

Exposed: use face fix back flap type hinges (26) for added security substitute one screw per leaf for a small bolt and nut, or use *clutch headed screws*.

Lid (16) (SW, Plywood, Chipboard, or MDF board)

The size and position of overhang is to suit the operator. Use one and a half pairs of *uncranked* 50 mm steel butts (24) to hinge the lid to the top edge of the tray. A hinge restraint or auto-stay should be used to prevent the hinges from becoming strained.

Note: if the top is to be used as a sawing board, the hinged edge should be positioned so that it faces the operative. This should reduce the risk of the top lifting during sawing operations.

Box interior The tool rack(14) should be detachable, and secured by two screws ($\frac{3}{4}$" × 8

Rd). Holes and slots may be cut to suit individual needs. A narrow sliding shelf or tray could be incorporated across the large compartment to slide between the plastic channel or plastic drawer runners.

As already mentioned, turn buttons (Fig. 3.2(a)) will be required for saw handles. Other useful attachments include a hanger (bracket) (Fig. 3.2(c)) to carry either a G cramp or coping saw and a bit holder (Fig. 3.2(d)). Both could be screwed either to the sides of the box ends or the partition.

Finish Fill with *stopper* all nail holes, screw holes, and blemishes etc. Sand down to receive the necessary coats of paint, or varnish of your choice.

Personalising It is traditional that the owner's name or initials should be visible for all to see. Lettering does not have to be hand-written; you could use a stencil or stick-on lettering. For all those who take pride in their work, I am sure this tradition will prevail.

Fig. 3.3 Porterbox with add-on (caddy, case and dolly) detached

Security (hasp and staple (27))   Fix the hasp to the underside of the lid, and staple to the face of the flap with through bolts and nuts. Link together by means of a padlock.

Just as all your tools should be individually marked, don't forget to also mark your toolbox with your name, post code, and/or telephone number – marks can be etched or punched into the wood on the inside or underside (preferably both) of the box.

Mobility   It is advisable to add two side handles (chest type (28)) to assist in lifting, especially if the top is to have an overhang, but because of the lever action of this type of handle, a 22 × 22 mm softwood section should be glued and screwed down the lower portion of all four corners, between ends (1) and front facing (17), ends (1) and back (23) to reinforce these corners. Two people can then handle the lifting better than one, particularly if it is to be longer than specified (to suit the length of your handsaw).

Castor base   All that is required is a solid or framed base with two, or four, strong surface fixing castors. In the former case movement would be achieved by lift and pull (or push). Case fasteners (toggles) are used to attach or detach the box from the base as and when required. If the worktop is to be used while the castor-base is in place, then brake boards under one or both ends (as shown in Fig. 3.2(b)) – as the case may be – can be used to restrict rolling.

Note: If the attachments (*Portercaddy/case*) are to be made, then the trolley (*Portertrolley*) will provide the castor base.

Figure 3.3 shows the *Porterbox* with the caddy, case and dolly detached.

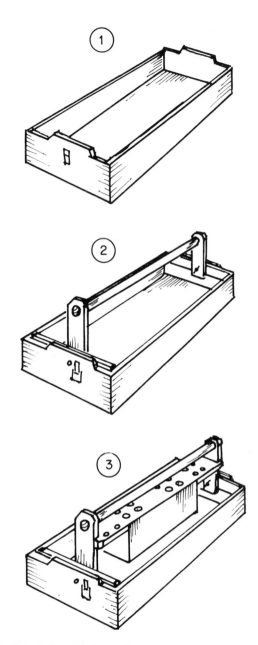

Fig. 3.4 Portercaddy tray options

Practical project 2: Portercaddy (Fig. 3.4)

This add-on attachment in the form of a tray provides extra base storage, whilst at the same time increasing the working height of the *Porterbox* worktop. With the addition of a handle the tray becomes a caddy – useful for transporting small quantities of tools etc, in an on-side situation (Fig. 3.4).

As can be seen from Fig. 3.6, provision has been made for a detachable tool rack to be incorporated in the construction – when not in use the rack can be used on top of the workbench. The folding bar carrying handle is retained within the tray when attached to the toolbox.

The *Portercaddy* is attached and detached by the use of heavy duty quick release toggles, or heavy duty trunk case clasps.

Sectional details of construction are shown in Fig. 3.5. A full cutting list is given in Table 3.2, in which components are named and numbered to coincide with the drawing.

Method and sequence of construction (Fig. 3.6)

Rod   Using the rod used to construct your *Porterbox*, extend the sections to accommodate all the attachments, taking extra care when allowing for clearance about the turning circles for the bar handle and tool rack as indicated in Fig. 3.5. This will permit both handle and tool rack to rotate freely through 180°.

Note: time spent at this stage of the proceedings usually means time and materials saved later on.

Table 3.2   Cutting list for Portercaddy, Portercase, and Porterdolly

Drawing no.	Item description	No. off	Material	L	W	T	Remarks
Tray							
1	End	2	Plywood	300	107	12	Multi-ply-form key
2	Sides	2	SW	676	95	18	
3	Bearers	2	SW	300	46	12	
4	Tray bottom	1	Plywood	676	315	4	
5	Locating blocks	4	Plywood	75	50	10	Also act as feet
Caddy tool rack							
6	Hanger	2	Plywood	215	50	12	
7	Handle	1	HW/SW	630		25 dia	Dowel/Broom handle
8	Peg-anti swivel	2	HW	50		6φ	Dowel
Caddy tool rack							
9	Shelf	1	SW	626	95	12	Holes to suit
10	Pelmet ends	2	SW	125	95	12	
11	Pelmet face	2	PL/Hdbd	300	137	3	
12	Hanger cleat	2	Plywood	64	50	12	
13	Shelf bearer	2	SW	50	22	12	
Case lid							
14	Frame	2	SW	628	45	12	Joints optional
15	Frame	2	SW	300	45	12	
16	Lock block	1	SW	75	45	12	
17	Top	1	PLY/Hdbd	628	270	3	
Dolly							
18	Bearer	2	SW	430	70	25	Support castors
19(a)	Rails (solid)	2	SW	540	45	34	
19(b)	Rails (laminated)	2	SW	540	45	25	
		2	SW	540	45	12	
20	Locating strip	2	SW	215	12	25	
Hardware							
21	Coachbolts	2		50		69	Washer/wingnut (2)
22	Toggles*	2					
23	Hinges (cranked)	2/3					Cabinet hinges, non-recessed type
24	Cupboard lock	1					
25	Case handle	1					
26	Castors - heavy duty	4					Base plate type

Key   SW - Softwood eg., redwood, whitewood or similar species (± 1 mm)   HW - Hardwood; straight grained species (± 1 mm)   PLY - Plywood; exterior quality not necessary   Hdbd - Hardboard (good quality); less expensive substitute for plywood   * - Strong lockable case trunk fasteners(clasp) can be used.

Tray

Ends (1) (12 mm Plywood)   Cut to shape (allow for keyed upstand). Then, while temporarily tacked together (face sides outermost), plane face edges, dress ends square and form the tool box locating key upstands.

Sides (2)   Cut to length. Check face sides (outside faces) for twist – remove any twist by planing. Temporary tack together (face sides outermost) and plane face edges. Then dress ends square and true to length – separate sides.

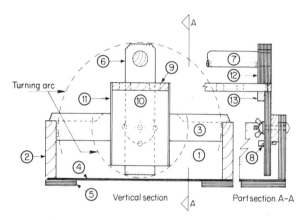

Fig. 3.5 Portercaddy - section details

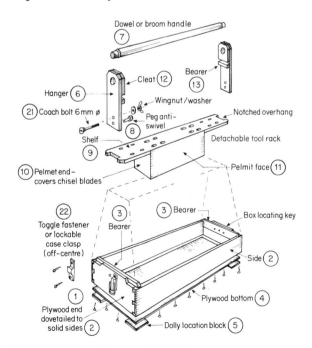

Fig. 3.6 Portercaddy - assembly details

Corner joint   Double-through or lapped dovetailed joints are recommended. Dovetails should be formed in the plywood ends (1). When cut, the ends can be separated. Progress to marking and cutting pins in the sides (2).

Keyway   Using the formed locating keys(1) as a template, carefully mark out and label the matching keyways onto the base of the Porterbox ends. Cut out the keyways, allowing only the very minimum of play. It is advisable to code these ends so that they match when the tray is assembled.

Bearers (3)   Prepare these to be in line with the top of locating keyed upstands (12 mm above tray ends), and notched over both sides.

Tray bottom (4) (4 mm Plywood)   Cut to size. Plane two edges to form one square corner.

Location blocks (5) (10 mm Plywood)   Cut and plane to size.

Tray assembly

1   Sand all inside faces of ends, sides, and bottom.
2   Glue and screw bearers in line with top of keyed upstand to plywood ends.
3   Glue and make all corner joints (cramp as necessary). Square framework by positioning square corner of plywood over frame. Glue underside of tray, tack bottom in position.

Note: At this stage, position the end keys into the box keyways, and check that the tray is square with it – adjust as necessary.

4   When alignment is true, screw ($\frac{3}{4}''$ × 8 Csk) onto sides and ends.
5   When set, dress all joints and trim bottom all round.
6   Attach *Porterdolly* locating blocks to each corner with glue and screws (1″ × 8 Csk).

Caddy

Hangers (6) (12 mm Plywood)   Cut to size, tack together, then plane to shape. Bore a series of holes to receive

1   shoulptered handle (25 mm Dia),
2   coach bolt (6 mm Dia),
3   anti-swivel and locking peg (6 mm Dia).

Handle (7)   Cut dowel to length. Reduce diameter of both ends to 22 mm by forming shoulder 12 mm in from each end.

Pegs (8)   Prepare from 6 mm and 12 mm dowel.

Assembly   Transfer, and bore holes through and/or into bearer and ends to take coach bolt and peg holes. Glue handle to hanger. Position coach bolt with washer and wingnut on the inside of tray. Anti-swivel pegs should be a push fit.

Caddy tool rack

Shelf (9)   Cut to length. Prepare by cutting notches for hangers and holes for tools.

Pelmet ends (10)   Cut to length and square ends. Glue and nail to underside of shelf.

Pelmet face (11) (3 mm Plywood)   Cut to size. Glue in squarely to pelmet ends and shelf, trim the edges and sand the whole assembly.

Hanger cleats (12) (12 mm Plywood)   Cut to size. Tack together, shape and bore holes with hangers. Glue and nail one to each hanger.

Shelf bearers (13)   Cut to length. Glue and screw ($\frac{3}{4}$″ × 6 Csk) to the hangers. Dress ends square with hanger edges.
   Note: When a detachable caddy tool rack is to be an option, hanger cleats are used. The bar handle is left loose to rotate within the widened hanger.

Attachment   Tray is held to the box by two toggles (22) (one at each end) or heavy duty trunk case clasps. If security is important, types with a locking eye are available – alternatively a small hasp and staple could be used.

Practical project 3: Portercase (Fig. 3.7)

By using the same tray construction as the *Portercaddy* with the addition of a hinged lid (fixed or detachable), the base unit can be transformed into a useful toolcase as shown in Fig. 3.7. An additional cutting list is shown in Table 3.2.

Method and sequence of construction (Fig. 3.8)

Rod   As the tray, but with the addition of the lid.

Lid framework   Choose a suitable corner joint for members (14) and (15). For example

a)   mortise and tenon (through or stopped),
b)   halving joint,
c)   dowelled,
d)   bridle.

   A butt joint could be used, with a triangular plywood gusset glued and pinned over the underside of each corner. Alternatively, a strip of plywood fixed along the full length of the hinged side of the lid would then leave a useful pocket within the underside of the lid.
   Note: before gluing joints together, do not forget to clean-up the inside edges, and to check that framework is square with the base tray. After the joints are made, dress and sand the faces before applying the top.

Top (17)   Cut to size, leaving a margin all round.

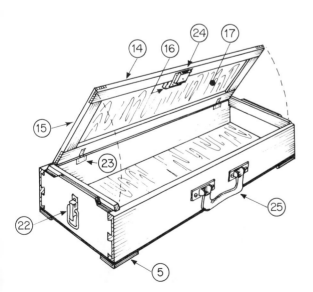

Fig. 3.7 Portercase

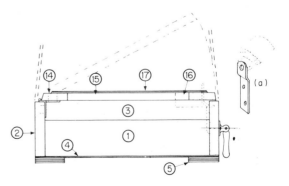

Fig. 3.8 Portercase – section details

Plane and bullnose edges. Pin to pre-glued upper face of framework.

Hinging lid (23)   Use three 50 mm non-recess cranked cabinet hinges.

Note: if the case is to be used in conjunction with the caddy, then consider using the lift-off type of cranked hinge.

Security   Alternatives include

a)   cupboard lock(24) fixed to lid.
b)   Purpose-made locking bar (Fig. 3.8(a)) – screwed to the inside of box, and passing through a mortise hole in the lid, allowing enough projection to enable a padlock to be inserted through the hole in the bar.

Mobility (25)   The case handle is fixed to the side with through-bolts.

Practical project 4: Porterdolly (Fig. 3.9)

The *Porterdolly* provides easy mobility via a castored base board – a useful 'add-on' for the workshop or wherever the floor is reasonably level. Construction and assembly details are shown in Fig. 3.9. The cutting list is given in Table 3.2.

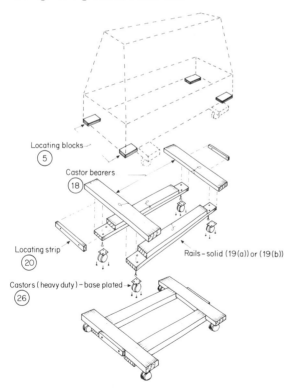

Fig. 3.9 Porterdolly - assembly details

Method and sequence of construction

Frame   Using the underside of the tray as a rod, cut bearers(18) and rails(19) to length. Check face sides for twist. Remove any twist by planing.

Solid rails (19a)   Cut away ends of rails to form lap-over bearers. Clean up all surfaces before making the joints with glue and securing with screws (1″ × 8 Csk) from the underside.

Laminated rails (19b)   Prepare the timber as above – butt rails up to bearers (face sides downward). Glue and screw laminating piece along rails and across the bearers to form the joints.

Locating strip (20)   With the caddy/case tray turned upside down, position the dolly frame within the location blocks. Cut the location strip to fit. Glue and nail the strip to the bearers.

Castors (26)   Screw heavy-duty base-plated castors to the underside of each bearer overhang.

Note: the overhang is to provide the box with greater mobil stability.

It is just as important to pay attention to

1   finish,
2   personalising,
3   security.

as you should have done with your *Porterbox*.

Figure 3.10 shows the *Porterbox* with the caddy, case and dolly attached.

Fig. 3.10 Porterbox with add-ons (caddy, case and dolly) attached

Practical project 5: Porterchest (Fig. 3.16)

This was originally designed to house various items of accompanying hardware to complement the main tool bit onsite such as:

1 Fixing devices;
   a) nails,
   b) screws,
   c) nut and bolts,
   d) knock down fittings (KDF),
   e) wall plugs,
   f) metal mending plates etc..
2 Window hardware;
   a) hinges,
   b) stays and fasteners,
   c) security locks and latches etc..

Fig. 3.11 Porterchest

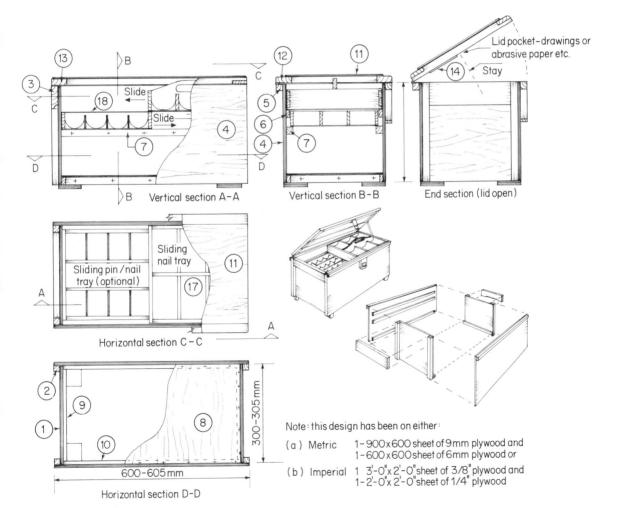

Fig. 3.12 Working drawings for Porterchest
Note:
1 All joints glued (PVA adhesive) - plywood to cleats, stiffeners and bearers to be screwed, bottom and lid facing nailed to bearers/framework cts mlt 75 mm
2 Hinges - 1 pair 60 mm steel or brass butts
3 Trays - compartments optional. Easy scoop bottoms may be formed with PVC 75 mm or 100 mm gutter

Note: this design has been on either:
(a) Metric   1 - 900 x 600 sheet of 9mm plywood and
           1 - 600 x 600 sheet of 6mm plywood or
(b) Imperial 1 3'-0"x 2'-0" sheet of 3/8" plywood and
           1 - 2'-0"x 2'-0" sheet of 1/4" plywood

3   Door hardware;
   a)   hinges,
   b)   locks and latches,
   c)   closers,
   d)   furniture (items of hardware fixed to the faces of the door and/or doorframe) i.e. lever handles, knobs, letter and finger plates etc.,
   e)   security (bolts, chains, alarms, viewers, etc).

It could, however, be used just as well as a small tool chest or a container for portable powered hand tools together with all their accessories, such as cutters, bits, blades, keys and spanners.

Its main features are its strength-to-weight ratio, its easy, economical method of construction, and simple yet effective means of security.

Its simple yet robust construction is based upon two small sheets of plywood (Fig. 3.1):

Metric: 900 × 600 mm sheet of 9 mm plywood,
   600 × 600 mm sheet of 6 mm plywood,
Imperial: 3′0″ × 2′0″ sheet of $\frac{3}{8}$″ plywood,
   2′0″ × 2′0″ sheet of $\frac{1}{4}$″ plywood,

and a glued and screwed softwood framework.

The interior of the chest is divided into compartments in the following way:

Base compartment – full length and width of the chest. The height will depend on the number and depth of the trays.

Lower sliding tray – half the area of the chest base, 45 mm deep, and sub-divided into a number of compartments.

Upper sliding tray (with liftout handle) – as lower tray but 70 mm deep.

Lid compartment – envelope to house drawing etc.

Means of mobility are provided by handholds located at each end of the chest, and a castor base to enable easy push and pull. With this provision the fully loaded chest should be easily manipulated in and out of those awkward situations such as under the work bench.

Method and sequence of construction

Figure 3.12 shows sectional working details of the *Porterchest*. The only modification to Fig. 3.12 is the size and positioning of the tray compartments.

A cutting list is shown in Table 3.3, in which all the various component parts are numbered. These numbers coincide with those on the drawings, and

Table 3.3   Cutting list for Porterchest

Drawing No.	Item description	No. off	Material	Finished L	Sizes (mm) W	(mm) T	Remarks
1	End	2	Plywood	305	305	9	
2	End cleats	4	SW	305	20	20	
3	End bearers	2	SW	265	20	70	Handhold
4	Front and back	2	PLY	605	305	9	
5	Stiffeners	2	SW	605	20	20	Hinge support
6	Tray runner	2	HW/SW	547	12	22	
7	Tray runner	2	HW/SW	547	20	20	
8	Bottom	1	PLY	547	305	6	
9	Bottom bearer	2	SW	305	12	22	
10	Bottom bearer	2	SW	547	12	22	
11	Lid top	1	PLY/Hdbd	605	305	4-6	
12	Lid frame	2	SW	605	45	18	
13	Lid frame	2	SW	363	45	18	
14	Lid pocket	1	PLY	547	100	4-6	
15	Staple/backing	1	SW	100	20	70	Cleat
16	Feet	4	PLY	75	75	12	Castor blocks

Sliding trays
17   Tray (70 mm deep) - suitable for large nails/screws/bolts etc.
18   Tray (45 mm deep) - suitable for small nails/screws/bolts etc.

Note: tray compartments to suit requirements.

Hardware							
19	Butt hinges	2	Steel/Brass	60			
20	Hasp and staple	1	Steel	100			
21	Castors	4	Steel/Plastics				Use with padlock Base
22	Auto-stay	1	Steel/Plastics				type (lid restraint)

Key   SW - Softwood eg. redwood, whitewood or similar species (± 1 mm)   HW - Hardwood; straight grained species (± 1 mm)   PLY - Plywood; exterior quality not necessary   Hdbd - Hardboard (good quality); less expensive substitute for plywood.

should therefore be used in conjunction with the following sequence. Note: As you progress with the sequence of construction, always bear in mind that all inside surfaces and edges should be cleaned up (planing, scraping, sanding, etc.) before being attached to another component.

Rod   Onto a large sheet of drawing paper, or piece of lining or wallpaper, attached onto a sheet of board material (for example, plywood or

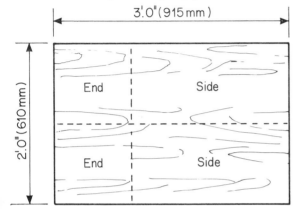

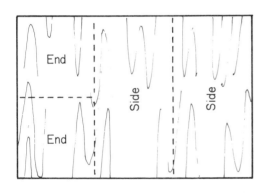

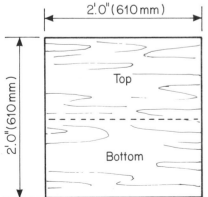

Fig. 3.13 Method of cutting the plywood sheets

hardboard) with adhesive tape, draw at least two full size sections through the chest. For example:

a)   a vertical lengthwise section (sectional front elevation),

b)   a vertical section through B–B.

The above details should include the trays and their compartments – it is at this stage that the tray depth, size, and number of compartments is decided upon. This process of drawing these full size details is known as *setting-out*.

Note: before setting out any rod, always check the sectional size of the available material, and, providing the tolerance is acceptable, these are the sizes which should be drawn onto the rod, *not those stated on the cutting list.*

Once the rod is complete in every detail (double check all the sizes, but these need not be shown on the rod), all future measurements i.e. lengths shapes etc. are taken by laying the material over the rod. Then, with a sharp pencil transferring any length marks etc., onto the material. This process is known as *marking-off* or, *marking-out*.

Not only does the method of using a rod save time, it greatly cuts down the risk of mistakes, thereby saving on materials.

Ends (1) (Plywood) (Fig. 3.13)   Cut to size, then, while still tacked together, plane the edges and corners square. Mark the face side and face edge and mark-off from the rod the position of bearers and cleats. Then separate.
After preparing the

> end cleats(2),
> end bearers(3) – not forgetting the cut-out for the hand hold,
> bottom bearers(9),

fix plywood to them using glue and screws ($\frac{3}{4}$″ × 8 Csk).

Front and back (4) (9 mm Plywood)   (Fig. 3.13). Prepare as the Ends **(1)**, then separate. After preparing the

> stiffeners,
> tray runners,
> bottom bearers,
> hasp and staple backing cleat(15),

fix to front and back panels (with the exception of **(15)** which is fixed to the front panel) using glue and screws ($\frac{3}{4}$″ × 8 Csk). It would be advisable when fixing tray runners to first tack them onto the plywood with a couple of panel pins before screwing through the plywood into the runners: In this way you are assured of their accurate positioning.

Bottom (8) (6 mm Plywood)   Cut and plane to size, making sure *all* corners are square.

Carcase assembly

1   Clean up all inside faces.
2   Pre-drill all the fixing points in the front, back, and bottom panels.
3   Glue and screw (1″ × 8 Csk) one corner together.
4   Position bottom panel up against the inside corner (as shown in Fig. 3.11) onto pre-glued bearers. When square with the sides, secure with screws ($\frac{3}{4}$″ × 6 Csk) to the bearers.
5   Continue to joint the remaining corners, and fix the bottom panel.
6   Reinforce each outer top corner by screwing (2″ × 8 Csk) though the ends of the front and back stiffeners into each end bearer.
7   Fix feet (castor blocks) to each corner using glue and screws (1″ × 8 Csk).

Lid

1   Prepare framework material (**12** and **13**), not forgetting to remove any twist.
2   Decide on a suitable corner framing joint, for example:
    a)   mortise and tenon or
    b)   halving joint.

    Make the joint.
3   Dress the joints as necessary with a smoothing plane, then sand both faces.
4   Cut and plane lid top(11) to size. Bullnose both long edges.
5   Glue and pin ($\frac{3}{4}$″ panel pins punched below

surface) lid top to framework.
6   Set to one side.

Note: if a lid compartment is to be incorporated (not shown in the photograph), then it can be attached at this stage.

Trays (17 and 18)   As mentioned earlier, the tray arrangement will depend upon the end use of the chest. I have however listed below a few suggestions:

1   12 mm thick material for the tray sides.
2   9 mm thick material for the tray divisions, with the exception of the handle section, which should be of 12 mm material.
3   The handle section should be made from either a slow grown softwood (redwood or whitewood) or a close grained hardwood because of shortness of the grain left at the ends of the slot cut out for the hand hold. Alternatively, and probably the most suitable, 12 mm birch plywood (multi-ply) could be used.
4   Corner joints should be dovetailed and glued together to offer the best resistance to any damage as a result of misuse.
5   Tray divisions should be housed together and into tray sides.
6   Tray bottoms (plywood or hardboard) should be glued and screwed ($\frac{3}{4}$″ × 6 Csk) to sides and glued and pinned ($\frac{5}{8}$″) to divisions.

Figure 3.14 shows the trays ready for assembly, whereas in Fig. 3.15 the assembled trays are fitted inside the chest but the tray bottoms are yet to be fixed.

Hinging lid.

1   Fit the lid to the chest top by planing.
2   Recess hinges (**19**) as necessary.

Fig. 3.14   Trays ready for assembly

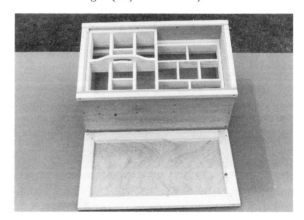

Fig. 3.15   Fitting trays into chest

3 Screw ($\frac{3}{4}''$ × 6 Csk) hinges to lid, then lid to chest.

4 Fix auto-stay (22) – lid packing may be required for the bracket, and a runner stop to prevent the trays impeding the movement of the stay arm.

Finish Fill with *stopper* all nail holes, screw holes, and blemishes etc.. Sand down to receive the necessary coats of paint or varnish of your choice.

Personalising As with the *Porterbox*, it is traditional that the owners name or initials should be visible for all to see. Lettering does not have to be hand-written; you could use a stencil, or stick-on lettering. For all those who take pride in their work, I am sure this tradition will prevail.

Security (Hasp and Staple (20)) Fix the hasp to the underside of the lid and staple it to the backing cleat with through bolts and nuts. Link together by means of a padlock.

Just as all your tools should be individually marked, don't forget to also mark your *Porterchest* with your name, post-code, and/or telephone number – marks can be etched or punched into the wood on the inside or underside (preferably both) of the chest.

Mobility I would strongly advise fixing castors (21) to the feet of the chest. These will make life a lot easier as the chest begins to fill up. The

completed chest with castors is shown in Fig. 3.16.

One final note – as the trays are filled sliding may become and difficult due to their increased weight. This movement can be made easier if candle wax is applied to the tray runners. Alternatively, purpose-made plastic runners could be used.

Fig. 3.16 Porterchest complete

4

Portable powered hand tools

Portable electric powered hand tools are available to carry out the following functions;

a) drilling,
b) sawing,
c) planing,
d) rebating,
e) grooving,
f) forming moulds,
g) screw driving,
h) sanding.

Provided a suitable supply of electricity is within easy reach, and the amount of work warrants their use, these tools can be used to speed up hand operations and where the use of permanent machinery would be impracticable.

Some portable power tools are produced specifically for the 'do-it-yourself' market and should not be confused with industrial tools – although they are principally the same, differences occur in both cost and the inability of do-it-yourself tools to sustain constant industrial use. For example:

Type	Category	Use
Light duty	Do-it-yourself	Occasional
Medium duty	General-purpose: tradesman	Moderate to intermittent
Heavy duty	Industrial	Continuous

This volume will deal with the following electrically driven tools:

1 drill,
2 rotary impact (percussion) drill,
3 screwdriver,
4 belt sander,
5 orbital sander.

Carpentry and Joinery 2 will deal with

1 the circular saw,
2 the reciprocating saw,
3 the planer,
4 the router,
5 the use of cartridge-operated fixing tools.

4.1 Electric drills (Fig. 4.1)

Choice of an electric drill will largely depend on

a) the type of work,
b) the volume of work,
c) the size of hole,
d) the type of material.

Figure 4.1(a) shows a 'Kango' palmgrip drill. Figure 4.1(b) shows the same drill being used to bore a hole in soft masonry (eye protection must be worn during such operations). Figure 4.1(c) illustrates a back-handle four-speed drill in use.

Ideally, an electric drill should be adjusted to rotate at a speed to suit both the material (workpiece) and the hole size (drill bit). Provided the drill is powerful enough and the correct drill bit (see Table 2.3) is used, it is possible to bore holes in most materials (including soft masonry).

A drill bit should rotate at its most effective cutting speed, otherwise it could overheat, quickly become dull, or even break. Cutting speed is often misquoted as revolutions per minute (rev/min),

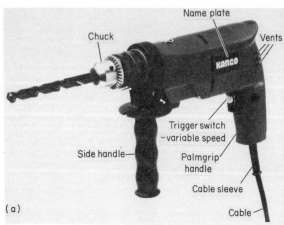

Fig. 4.1(a) Palmgrip electric drill

Fig. 4.1(b) Palmgrip electric drill in use

Fig. 4.1(c) Back handle electric drill in use

which only denotes the number of times the chuck revolves every minute (the drill speed). To determine a drill bit's cutting speed (edge speed of bit), the distance it travels every revolution must be known. As can be seen from Fig. 4.2, the distance covered by one revolution will vary according to the diameter. Therefore to find the cutting speed of a drill bit – which is usually quoted as metres per second (m/s) – we use the following formula:

$$\text{Cutting speed (m/s)} = \pi \times \text{Diameter (m)} \times \text{Drill speed (rev/s)}$$

where $\qquad \pi(\text{'pi'}) = 22/7 \text{ or } 3.142.$

For example, for a drill bit of 6 mm diameter in a drill with a working speed of 1500 rev/min,

$$\text{Cutting speed} = \pi \times \text{Diameter} \times \text{Drill speed}$$

where $\quad \text{Diameter} = 6 \text{ mm} = \dfrac{6}{1000} \text{ m}$

and $\quad \text{Drill speed} = 1500 \text{ rev/min} = \dfrac{1500}{60} \text{ rev/s}$

$\therefore \quad \text{Cutting speed} = \dfrac{22}{7} \times \dfrac{6}{1000} \times \dfrac{1500}{60} \text{ m/s}$

$\qquad\qquad\qquad = 0.47 \text{ m/s}.$

Drill bits should be used within the recommended ranges of cutting speeds in Table 4.1.

Drilling holes in wood usually requires about twice the cutting speed for metal, but, because of the many hard and abrasive wood-based materials now in common use, consideration should be given to the material's composition before choosing a bit or its speed.

Table 4.1 Recommended cutting speeds

Material	Cutting speed (m/s)
Aluminium	1.00 to 1.25
Mild steel	0.40 to 0.50
Cast iron	0.20 to 0.40
Stainless steel	0.15 to 0.20

Drill manufacturers usually recommend the appropriate rev/min to suit both the drill bit size and the material. As a general guide, the larger the hole the lower the rev/min.

Method of use (Figs 4.1(b) and (c))

How the drill bit is applied to a workpiece will vary according to the workpiece's size and shape, but it is usually a two-handed operation. It is therefore essential that the work is securely held by a clamp or is such that it will not be affected by the pressure needed to drill the hole. Pressure should be regulated to allow the bit to cut into the material, to keep the hole clear of waste particles, and also to avoid breaking through the under side. Insufficient pressure could result in both the bit and workpiece being overheated, due to heat being generated by friction.

Attachments

Although most manufacturers supply a large range of drill attachments, only those which aid the drilling process should be considered suitable for use by the tradesman; for example,

a) an angle-drilling attachment (Fig. 4.3),
b) a drill stand (Fig. 4.4),
c) a mortiser (Fig. 4.5).

Figure 4.3 shows an angle-drilling attachment in use. This is ideal for getting into awkward corners or spaces between floor joists etc.

A drill stand permits the correct amount of pressure to be applied and leads to greater drilling accuracy. Figure 4.4 shows a bench drill stand being used with a chuck-and-drill-bit guard in

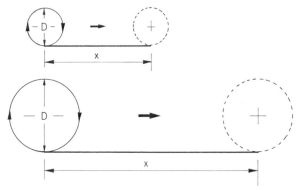

D = Diameter of drill bit

x = Distance covered by cutting periphery (one revolution)

Fig. 4.2   Drill bit cutting speed

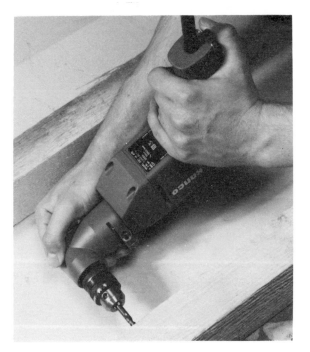

Fig. 4.3   Kango angle-drilling attachment

position. It is mandatory to fit and use such a guard to comply with the Factories Act 1961 and the Health and Safety at Work Act 1974.

The chisel mortiser works on a similar principle to a drill stand. A special bit drills a hole at the same time as the chisel is paring it square. Figure 4.5(a) labels all the main parts, and Fig. 4.5(b) shows it in use; note how the cable from the drill is kept well away from the work bench. This attachment is very effective – mortise holes of up to 13 mm × 13 mm can be cut every time the handle is brought down.

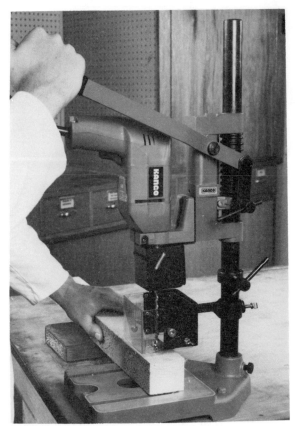

Fig. 4.4   Kango bench drill stand and drill

4.2 Rotary impact drills (Fig. 4.6(a))

These have a dual purpose: they can be used for rotary drilling or, by the turn of a knob or lever, they can be transformed into an impact drill which with the use of a special tungsten-carbide-tipped drill bit can bore holes into the hardest of masonry or concrete.

Figure 4.6(a) labels the main parts of a 'Kango' impact drill.

Palmgrip drills can be used for holes up to 10 mm in diameter. Figure 4.6(b) shows the drill in use – eye protection (snug-fitting goggles). Note that the operator is using snug-fitting full goggles, giving complete eye protection, which must be provided and used in such or similar situations to comply with the Protection of Eye Regulations 1974. N.B. Depth-stop attachments are available to limit the depth of the hole to suit a plug depth etc.

Back-handled machines allow much larger holes to be bored. Figure 4.6(c) shows such a machine in use.

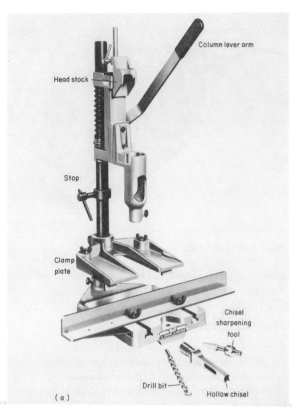

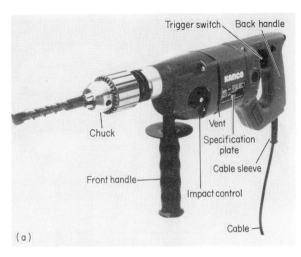

Fig. 4.6(a) Rotary impact drill

Fig. 4.6(b) Palmgrip impact drill in use

Fig. 4.6(c) Back handle impact drill in use

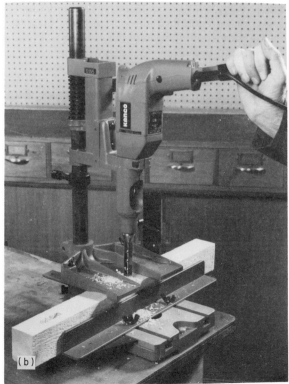

Fig. 4.5 Kango chisel mortiser and stand

4.3 Electric screwdriver

This offers a fast virtually effortless method of driving screws. Screwdriver bits of different sizes are available to suit both *slot* and *Pozidriv* (*Superdriv*) heads.

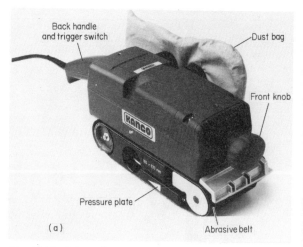

Fig. 4.8(a) Belt sander

Fig. 4.7 Electric screwdriver

Method of use

The screwdriver being used in Fig. 4.7 is one manufactured by Kango Ltd, all of which have a mechanism known as a positive clutch with depth-setting device, which enables a screw to be driven to a predetermined depth, after which the screwdriver bit stops turning. The larger screwdrivers are capable of driving woodscrews of up to 100 mm in length. It should be noted that both clearance and pilot holes (Fig. 15.6) should be pre-bored before driving any woodscrew.

4.4 Belt sander (Fig. 4.8)

This is designed and constructed to tackle heavy sanding problems with minimum effort. Figure 4.8(a) names the main parts of this machine. The endless abrasive belt is driven by a motor-driven rear (heel) roller over a front (toe) belt-tensioning roller and then over a steel-faced cork or rubber plate on its base – this is the part that makes

contact with the workpiece. Dust is discharged via a suction-induced exhaust into the dust bag.

Method of use

Figure 4.8(b) shows a belt sander being used. It should be permitted to reach full speed before being gently lowered on to the workpiece – allowing its 'heel' to make contact slightly before its 'toe', to avoid any kick-back. When contact is made, there will be a tendency for the sander to move forward and force the workpiece backwards, due to the gripping action of the sanding belt (see Fig. 4.9), so both the sander and workpiece must be held firmly at all times. The surface finish produced by the sander will depend on what grade of abrasive belt has been used. (See Section 2.9; Table 2.6 and Fig. 2.95).

Fig. 4.8(b) Belt sander in use

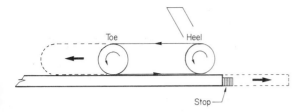

Fig. 4.9 Mechanical action of a belt sander

Fig. 4.10(b) Orbital sander in use

Note: the dust bag must remain attached to the sander while the motor is in motion, otherwise dust and particles of grit will be discharged directly from the exhaust tube at an alarming rate and could result in serious injury.

4.5 Orbital sander or finishing sander
(Fig. 4.10)

The main parts of this machine are illustrated in Fig. 4.10(a). The abrasive sheet is attached by clips or levers (depending on the design) to a felt or rubber sanding pad which is designed to orbit around a 3 mm diameter at about 12 000 rev/min.

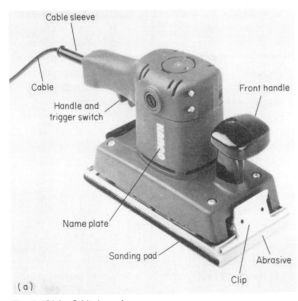

Fig. 4.10(a) Orbital sander

Method of use

Figure 4.10(b) shows an orbital sander being used. The self-weight of the sander is usually sufficient pressure – anything other than light pressure could result in scratching, clogging the paper, or even the body orbiting while the pad appears to remain stationary.

To obtain the best results, the appropriate grade of abrasive paper must be used to suit the job. This often means starting with a coarse grade of paper, then reducing the grade until the desired result is achieved (see Section 2.9, Table 2.6 and Fig. 2.95).

As with all mechanical-sanding operations, dust can be a serious health hazard, and, even though some orbital sanders have dust bags similar to belt sanders, mouth and nose masks will be required during some operations.

4.6 Specification plate (SP)

A label is fixed to the outer casing of all portable power tools, giving information under some, if not all, of the following headings:

a) *Manufacturer* – maker's name or trade mark.
b) *Type number* – provides a method of identification for attachments or spare parts.
c) *Capacity* – chuck opening size
   – rev/min for 'high' and 'low' speed ratings
d) *Voltage* (potential difference) – the supply must be within the range stated on the SP; for example '220/240 V – 105/130 V'. All portable power tools used on building and construction sites should be run from a 110 volt supply, or from higher-voltage supplies which have been suitably reduced via a step-down (centre-tapped) transformer to 110 volts (see Fig. 4.13). Reduced voltage greatly reduces the risk of an electric shock being fatal.

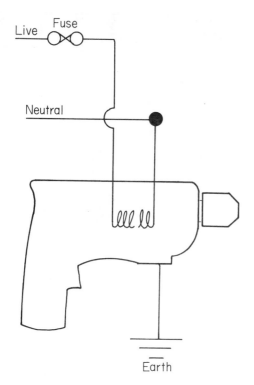

Fig. 4.11 Normal earthed system

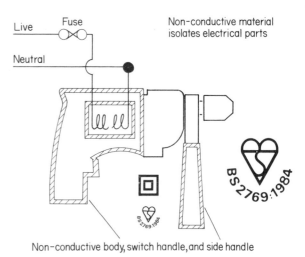

Fig. 4.12 Double insulated system

However, the risk can be further reduced by 'double insulation' (see below).

e) *Wattage* (input on full load) describes the power, or rate of electricity consumption by the motor, expressed in watts, where

$$\text{Watts} = \text{Volts} \times \text{Amperes}.$$

f) *Amperage* – the current taken by the motor, expressed in amperes. This can be related to power and voltage;

$$\text{Amperes} = \frac{\text{Watts}}{\text{Volts}}$$

or to voltage and resistance;

$$\text{Amperes} = \frac{\text{Voltage}}{\text{Resistance}}.$$

Therefore when the voltage is reduced from 240 V to 110 V the current must be increased if the same power is to be maintained.

g) *Electrical safety* – The Factories Act (Electrical Regulations) 1908–1944 protects the user of an electrical appliance by requiring that any metal part of that appliance which could come in contact with a current-conducting wire must be earthed.

Earthing and insulation

Figure 4.11 shows how an electric drill or similar machine is protected by an earth line. The principle is as follows. The neutral of the electrical supply is earthed at the power station or distribution station. If the electrical appliance has a low-resistance connection to earth, then a return path is available for the current if a fault occurs that causes the appliance's casing to become live. This low-resistance path allows a high current to flow which causes the fuse (providing it is correctly rated) to burn out (blow), thus stopping the flow of current and rendering the machine safe.

No matter how effective this system might seem, there is always the possibility that it will not work when it is most needed. For example, the system will fail to work if the earth wire has

1 not been connected to the plug socket, or has become defective en route or at source,
2 become disconnected from the machine,
3 become disconnected from the plug,
4 been damaged within the flexible cable to the machine, or the extension lead.

If the earthing system fails to work, possibly for one of the above reasons, it could result in the operator's body being used as an escape route to earth – the results of which could prove fatal.

Fortunately, nearly all portable power tools produced are now *double-insulated*. Figure 4.12 shows how a double barrier is formed around all those components capable of conducting an electrical current. This is achieved by using a strong non-conductive material for the body and/or

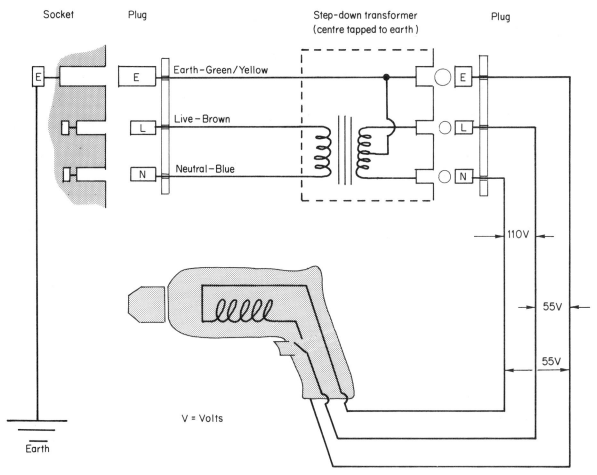

Fig. 4.13 110V system

isolating any metal parts with a non-conductive inner lining, thus eliminating the need for an earth wire. Portable power tools which are double-insulated bear the symbol of a *square in a square* (Fig. 4.12) and, before these tools can legally be used in industry, dispensing with their earth wire, they must comply with the Factories Act (Electrical Regulations) Portable Apparatus Exemption Order 1968 by being approved by the British Standards Institution and bearing BSI's 'Kitemark' BS 2769:1984 on the casing (Fig. 4.12).

Double-insulated tools are undoubtedly safer than single-insulated tools (normal earthed systems) but, unless they are powered from a low-voltage supply (110 V), there is still considerable danger from the current-carrying cable.

Figure 4.13 shows how the supply voltage of 240 volts is reduced to 110 volts, with a centrally tapped earth, so that, if a break-down in insulation does occur, the operative should only receive a shock from 55 volts.

Voltage-operated earth leakage circuit breakers (ELCB) These are devices which, in the event of an electrical fault occurring where current flows to earth, thereby putting the operative at risk of an electrical shock, disconnect both live and neutral supplies very quickly (only double pole ELCBs must be used). They do, however, depend on the proper functioning of their electrical and mechanical components. It is therefore essential that they are tested at regular intervals (before each work period) – test buttons are provided to enable this test to be carried out.

ELCBs are available in several forms; the type to be used with portable powered hand tools is one which is permanently combined with the plug and directly wired to the tool. Each tool should have its own ELCB (ELCB adaptors should not be used as there is always the risk that they may accidentally be forgotten or left out of the system).

4.7 General safety

Before a portable power tool is used, the operator must be confident that all necessary steps have been taken to ensure both his or her safety and that of any persons within close proximity of the operation to be carried out.

The precautions listed below should always be followed:

1 Never use a portable power tool until you have been instructed in its use by a competent person.
2 Only use a portable power tool after authorised approval (the tool in question may have been withdrawn from use for some reason of safety).
3 The manufacturer's handbook of instruction for the tool in question should be read and understood before use.
4 Always wear sensibly fitting clothes – avoid loose cuffs, ties, and tears etc.
5 Wear eye protection where required by the Protection of Eyes Regulations 1974.
6 Dust masks should be worn where the operative's health may be at risk.
7 Guards, where fitted, must always be used.
8 Never use blunt or damaged cutters.
9 Keep flexible cables away from the workpiece, cutters, and sharp edges and from trailing on the floor.
10 Before changing bits or abrasive sheets or making any adjustments, always disconnect the tool from the electric supply (remove the plug from its socket).
11 If a tool is damaged or found to be defective, return it to the stores or the person responsible for it. Ensure that it is correctly labelled as to the extent of its damage or defect.
12 If injury should occur – no matter how minor – first aid must be applied immediately to avoid the risk of further complications. The incident should then be reported to the person responsible for safety.

5

Woodworking machines

The aim of this chapter is to help the student to become aware of the more common types of woodworking machinery that the carpenter and joiner may encounter, to be able to recognise these machines by sight, and to understand the basic function of them. *Carpentry and Joinery 2* will discuss the use of these machines.

Undoubtedly the most important aspect of any woodworking machine is its safe use. To this effect, set rules and regulations are laid down by law and

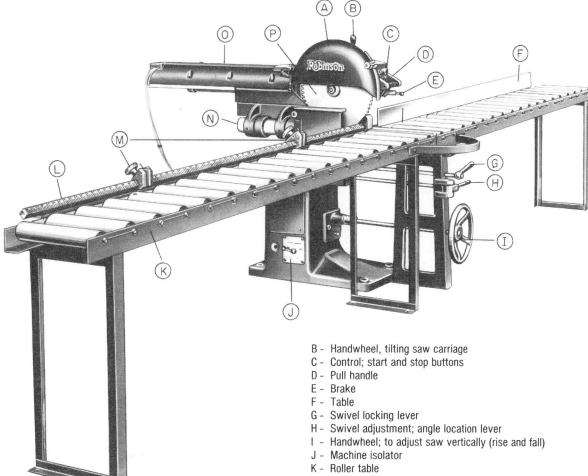

Fig. 5.1 Thomas Robinson's cross-cutting and trenching machine

A - Saw guard with adjustable front shield
B - Handwheel, tilting saw carriage
C - Control; start and stop buttons
D - Pull handle
E - Brake
F - Table
G - Swivel locking lever
H - Swivel adjustment; angle location lever
I - Handwheel; to adjust saw vertically (rise and fall)
J - Machine isolator
K - Roller table
L - Cutting-off gauge bar
M- Adjustable stops
N - Sawdust exhaust
O - Travelling carriage
P - Saw blade

must be carried out to the letter and enforced at all times. The need for such strict measure will become apparent – especially when one considers that, unlike in most other industries, the majority of our machines are still fed by hand, thus relying on the expertise of the skilled operator, who must concentrate and exercise extreme caution at all times.

5.1 Cross-cutting machines

These machines are designed to cut timber across its grain into predetermined lengths, with a straight, angled, or compound-angular (angled both ways) cut. It should be said, however, that their primary function is to cut long lengths of timber into more manageable lengths. With a standard blade and/or special cutters, they can also be used to cut a variety of wood joints, for example housing, halving, mitred, dovetailed, and birdsmouth joints.

There are two main types of cross-cut saws: the travelling-head and radial-arm types.

The travelling-head cross-cut saw (pull-over saw)

Figure 5.1 shows a typical saw of this type. The saw, which is driven direct from the motor, is attached to a carriage mounted on a track, which enables the whole unit to be drawn forward (using the pull handle) over the table to make its cut. The return movement is spring-assisted. The length of timber to be cut is supported by a wood or steel roller table and is held against the fence. For cutting repetitive lengths, a graduated rule with adjustable foldaway stops can be provided. Angle and height adjustments are made by operating the various handwheels and levers illustrated.

The radial-arm cross-cut saw

This carries out similar functions to the travelling-head cross-cut type but differs in its construction by being lighter and having its saw unit, together with its carriage, drawn over the workpiece while they are hung from under an arm which radiates over the table.

Some of these machines (like the one shown in Fig. 5.2) are generally more versatile than the travelling-head types, and what helps to make them so is that not only does the carriage arm swivel 45° either way but also the saw carriage tilts from vertical to horizontal and revolves through 360° – enabling ripping, grooving, and moulding operations to be carried out.

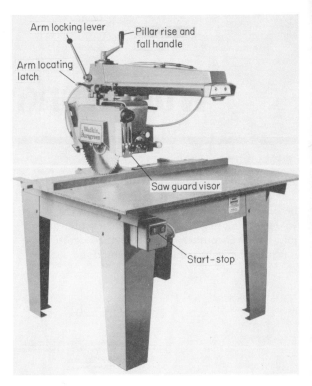

Fig. 5.2 Wadkin universal radial cross-cut saw

5.2 Circular sawing machines

For the purpose of this chapter, circular sawing machines will be taken to be those machines which have circular saw blades housed for their greater part inside a metal saw bench and are used to divide squared stock (rectangular- or square-sectioned timber).

Hand feed circular saw benches

These are primarily used for resawing timber lengthwise in its width (ripping or flatting) or its depth (deep-cutting or deeping). Some models can vary the depth at which the blade projects above the saw-bench table.

Figure 5.3 shows a typical general-purpose saw bench with provisions for a cross-cutting fence. The dotted area at the back of the saw bench indicates where a *backing-off* table must be positioned when anyone is employed to remove cut material from the delivery end (see Regulation 20(2), *Carpentry and Joinery 2*).

Saw benches of this type use relatively large saw blades which require *packings* to help prevent the saw blade deviating from its straight path. Packings are pieces of oil-soaked felt, leather, or

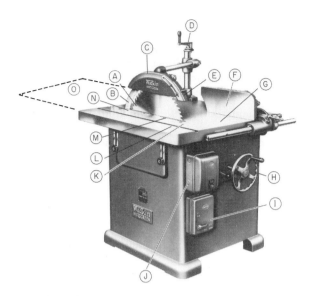

Fig. 5.3 Wadkin Bursgreen circular saw bench
A - Saw blade
B - Riving knife
C - Crown guard
D - Crown guard adjustment
E - Extension guard
F - Adjustable fence (will tilt up to 45°)
G - Table
H - Handwheel; saw spindle rise and fall
I - Isolator
J - Control; start and stop buttons
K - Mouthpiece (hardwood)
L - Machine groove for cross-cutting gauge
M- Saw blade packing
N - Finger plate (access to saw spindle)
O - Extension table (provision to comply with Regulation 20(2))

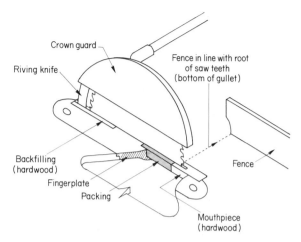

Fig. 5.4 Packing circular saws

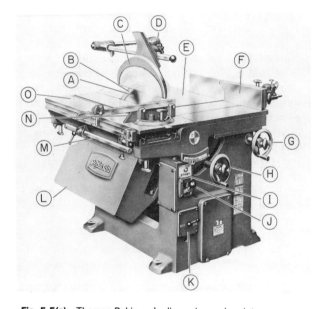

Fig. 5.5(a) Thomas Robinson's dimension and variety saw
A - Saw blade
B - Riving knife
C - Crown guard
D - Crown guard adjustment
E - Main table
F - Adjustable fence (will tilt up to 45°)
G - Handwheel; saw rise and fall
H - Handwheel; saw tilt adjustment
I - Control; start button
J - Combined brake and stop lever
K - Isolator
L - Tilting saw frame
M- Sliding table stop (adjustable)
N - Mitre and cross-cut fence and gauge
O - Sliding table (rolling)

similar materials, specially made by the wood-cutting machinist to suit the various types of blade and their relevant position above the saw-bench table. Packings, together with a hardwood *mouthpiece* and *backfilling*, can be seen in Fig. 5.4.

The mouthpiece acts as a packing stop and helps prevent the underside of sawn stock from splintering away. Like packings, it will be made to suit each size of saw blade. Backfilling protects the edges of the table from the saw's teeth – one side is fixed to the table; the other to the *fingerplate*. The fingerplate lifts out the table to facilitate changing a saw blade.

Dimension saw

This uses a smaller saw blade, thus limiting its maximum depth of cut to about 140 mm,

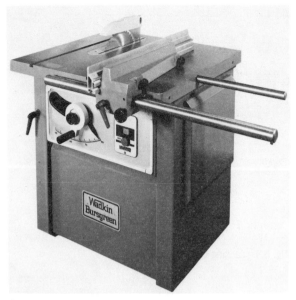

Fig. 5.5(b) Wadkin Bursgreen AGS 250 tilting arbor dimensioning saw

depending on the size of blade and the saw-bench capacity. Dimension saw benches like the one illustrated in Fig. 5.5(a) are capable of carrying out a variety of sawing operations with extreme accuracy and produce a sawn surface which almost gives the appearance of having been planed.

Sawing operations with this saw include ripping, deeping, cross-cutting, mitre and bevelled work, and – provided statutory guarding requirements are met – grooving and moulding. However, for work other than normal sawing operations (usually the most dangerous), effective guarding may not be practicable, in which case that type of work should not be done.

The main features which enable such a variety of operations to be done are;

1	adjustable double fence (tilt and length)	– adapts to suit both ripping and panel sawing,
2	cutting-off gauge	– allows straight lengths, angles, or single and double mitres to be cut,
3	tilting saw frame and arbor (main spindle)	– facilitates bevelled cutting etc.,
4	draw-out table	– for access to the saw arbor, for saw or cutter changing,
5	rolling table	– for panel cutting and squaring.

A more modern dimension saw is shown in Fig. 5.5(b).

Panel saw

Figure 5.6 shows a panel saw which is similar in design to the modern dimension saw, but having features dedicated to the cutting of sheet materials only. These features include

smaller diameter blade	– due to limited thickness of material to be cut,
crown guard attached to riving knife	– removing the need for a support pillar (as on the dimension saw) restricting the sheet size that can be cut.
extended sliding/rolling table	– to cope with full sheet size,
scoring saw (optional – see insert Fig. 5.6)	– a small diameter blade pre-cutting the underside of veneered panels, therefore preventing break out.

5.3 Planing machines

These machines are used to smooth the surface of the wood and reduce sawn timber to a finished size (see Fig. 1.19). The first operation – known as *flatting* – must produce a face side which is straight, flat, and twist-free. This is followed by straightening a face edge which must be square to the face side – known as *edging*. The timber is then reduced to thickness by planing the opposite faces parallel throughout their length.

Hand feed planer and surfacer

This is used for planing the face side and face edge. Figure 5.7(a) illustrates a typical traditional surfacing machine; Fig. 5.7(b) shows a more modern machine. Its long surfacing table supports the timber as it is passed from the infeed table, over the circular cutter block (see Fig. 5.8), to the outfeed table. In accordance with the regulations, cutter guards (bridge and back guards) must always be in place during all planing operations. The fence can be moved to any position across the table and be tilted for bevelled work.

Some surfacing machines include in their design facilities to carry out such functions as rebating and moulding. Operations such as these require special means of guarding the cutter and must not be carried out unless such means are provided.

Sliding table   Extension table

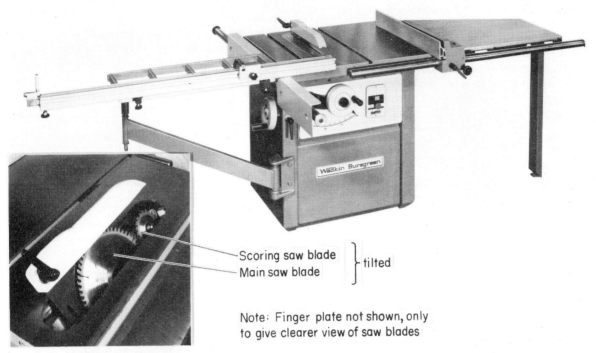

Scoring saw blade ⎫
Main saw blade ⎬ tilted

Note: Finger plate not shown, only to give clearer view of saw blades

Fig. 5.6 Wadkin Bursgreen AGSP panel saw bench

(a)

Fig. 5.7(a) Thomas Robinson's hand feed planer and surfacer
A - Pressure bars (holding down springs)
B - Adjustable fence (will tilt up to 45°)
C - Infeed table
D - Handwheel; infeed table height adjustment
E - Handwheel; infeed table lock
F - Main frame
G - Control; start and stop buttons
H - Handwheel; outfeed table lock
I - Side-support table (extra support while rebating)
J - Handwheel; outfeed table height adjustment
K - Outfeed table (delivery table)
L - Telescopic bridge guard

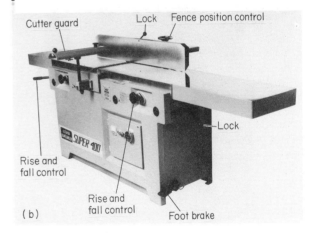

(b)

Fig. 5.7(b) Wadkin Bursgreen S400 surface planer

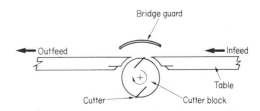

Fig. 5.8 Surfacer cutter - block arrangement

Thicknesser or panel planer

A thicknesser or panel planer such as the one illustrated in Fig. 5.9 is used for the final part of the planing process – reducing timber to its finished size.

After the table has been set to the required thickness, timber is placed into the infeed end where it is engaged by a serrated roller which drives it below a cutter block. The machined piece is then delivered from the other end by a smooth roller. Two anti-friction rollers set in the table prevent any drag. The arrangement is shown in Fig. 5.10.

Combined hand and power feed planer

This is one machine, capable of both surfacing and thicknessing. Figure 5.11(a) shows a traditional combined machine, Fig. 5.11(b) a more modern machine.

Unlike the purpose-made thicknesser, when these machines are used for thicknessing, that part of the cutter block which is exposed on the surfacing table must be effectively guarded throughout its length.

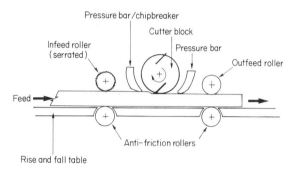

Fig. 5.10 Thicknesser table and cutter block arrangement

Fig. 5.9 Wadkin Bursgreen roller feed planer and thicknesser
A - Thicknessing table (infeed)
B - One of two anti-friction rollers
C - Cutter block guard and chip chute
D - Thickness scale
E - Control; start and stop buttons
F - Feed speed selector switch (4.5 m/min and 9 m/min)
G - Handwheel; raise and lower table
H - Lever; table lock
I - Outboard roller

Fig. 5.11(a) Thomas Robinson's hand and power feed planer
A - Pressure bars (holding-down springs)
B - Adjustable fence (will tilt up to 45°)
C - Infeed table (surfacing)
D - Surfacing table adjustment; raise and lower
E - Idle roller to thicknessing table
F - Surfacing infeed table lock
G - Main frame
H - Control; start and stop buttons
I - Handwheel; thicknessing table rise and fall
J - Surfacing outfeed table (delivery table) lock
K - Finished thickness scale
L - Telescopic bridge guard
M- Outfeed table (surfacing)

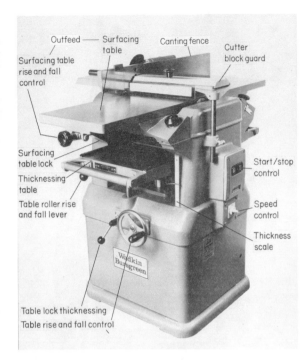

Fig. 5.11(b) Wadkin Bursgreen surface planer and thicknesser

5.4 Mortising machines

These machines cut square-sided holes or slots to accommodate a tenon. The hole is made either by a revolving auger bit inside a square tubular chisel or by an endless chain with cutters on the outer edge of each link. Machines are made to accommodate either method or a combination of both.

Hollow-chisel mortiser

Figure 5.12 shows the various components of this machine (see also Fig. 4.5). As the mortising head is lowered, the auger bores a hole while the chisel pares it square. The chippings are ejected from slots in the chisel side. After reaching the required depth – which for through-mortise holes is about two thirds the depth of the material – the procedure is repeated along the desired length of the mortise. The workpiece is then turned over and reversed end for end, to keep its face side against the fence, and the process is repeated to produce a through mortise (see Fig. 5.13).

Chain and chisel mortiser

Figure 5.14 shows a machine which is capable of mortising by chain as well as by hollow chisel.

The chain mechanism shown in Fig. 5.15

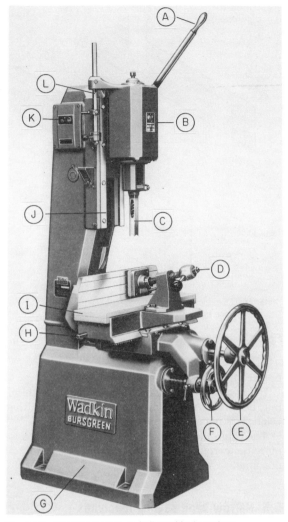

Fig. 5.12 Wadkin Bursgreen hollow chisel mortiser
A – Operating levers
B – Mortising head
C – Hollow chisel and auger
D – Clamp (faced with a wooden plate)
E – Handwheel; operates table longitudinal movement
F – Handwheel; operates table cross-traverse
G – Main frame
H – Table stop bar
I – Work table (timber packing) and rear face
J – Mortising head slideway
K – Control; start and stop buttons
L – Depth stop bar (mortise depth adjustment)

consists of the chain, a guide bar and wheel, and a sprocket which turns the chain at high speed.

The guide-bar section (fully guarded at all times) is lowered into the securely held workpiece to cut a large slot (depending on the chain size) in one operation of the lever arm. To prevent the chain splintering away the surface of the wood on its

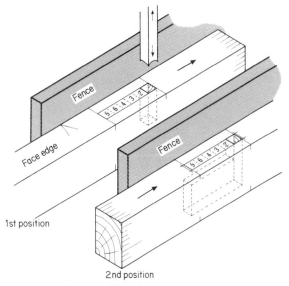

Note: (a) Clamp not shown
(b) Face side positioned towards fence

Fig. 5.13   Hollow chisel mortiser; cutting a mortise hole

upward motion, a chipbreaker is used.

Using a chain can be much quicker than a chisel but, because of the semicircular-bottomed hole that is left, it is not suitable for short or stub tenons.

The chain mortiser is regarded as being much more dangerous to use than the hollow-chisel machine.

5.5 Narrow band-sawing machines

Unlike those band-saws featured in Chapter 1, these machines have blades which do not exceed 50 mm in width. The band-saw illustrated in Fig. 5.16 has a maximum blade width of 38 mm.

These smaller machines are used for all kinds of sawing, from cutting freehand curves – the radius of which will depend on the blade width (see *Carpentry and Joinery 2*) – to ripping, deep-cutting, and cross-cutting. By tilting the table, angled and bevelled cuts can be made.

The machines consist of a main frame on which two large pulleys (wheels) are fixed. The bottom wheel is motor-driven, while the top wheel is driven by the belt action of the saw blade. The pulley (wheel) rims are covered with a rubber tyre to prevent the blade slipping and to protect its teeth.

To facilitate blade tensioning and alignment, the top wheel can be adjusted vertically and tilted sideways. The amount of tension will depend on

Fig. 5.14   Thomas Robinson's chain and chisel mortiser
A - Chain/chisel operating levers
B - Chisel headstock
C - Chisel mortising, depth stop arrangement
D - Hollow chisel and auger
E - Work table (timber packing) and rear face
F - Control reset button
G - Isolator
H - Handwheel; table rise and fall
I - Handwheel; operates table cross-traverse
J - Handwheel; operates table longitudinal movement
K - Clamp
L - Chipbreaker
M - Chain guard and window
N - Chain mortising, depth stop arrangement
O - Chain mortising headstock

the blade width – incorrect tension could result in the blade breaking.

Saw guides give side support to the blade above and below the table while cutting takes place. Back movement of the blade is resisted by a *thrust wheel*

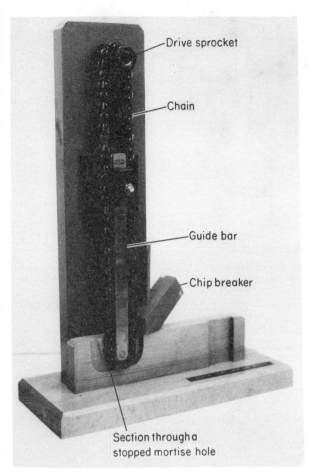

Fig. 5.15 Model showing the arrangement of a mortise chain, its mechanism and application

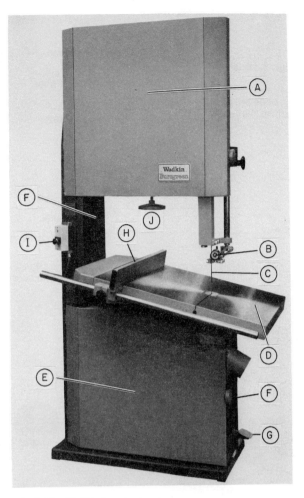

Fig. 5.16 Wadkin C series bandsaw
A - Upper swing-away guard/door
B - Saw guides and thrust wheel (adjustable guard)
C - Band saw blade
D - Table (shown tilted)
E - Lower swing-away guard/door
F - Main frame
G - Foot brake
H - Fence
I - Control; start and stop buttons
J - Handwheel; for regulating saw tension

or disc which revolves whenever the blade makes contact. The position of the guides and thrust wheel relative to the blade is critical if efficient support to the blade is to be maintained. To this end, they are provided with a range of adjustments to allow for varying blade widths. The assembly of guides, thrust wheel, and blade guard adjusts vertically so that it can be positioned as close as practicable to the workpiece (see Fig. 5.17).

5.6 Wood-turning lathes

Most students will have seen, if not used, a basic wood-turning lathe during their school days. It was, and still is, a very popular way of introducing students to one of the less dangerous woodworking machines. It is designed to rotate a piece of solid wood while the operator uses a chisel to form it into a round or cylindrical shape.

Wood-turning lathes have changed very little over the years, except that they have become safer to use. There is, however, still a danger of the workpiece working loose and of articles of clothing etc. becoming caught in unguarded moving parts.

Many modern lathes, such as the one shown in Fig. 5.18, include in their design such features as simple speed control and a spindle brake. Extra versatility can be given to this machine by attaching a travelling carriage and toolpost to the lathe bed, which allows accurate uniform cuts to be made laterally and transversely in both wood and

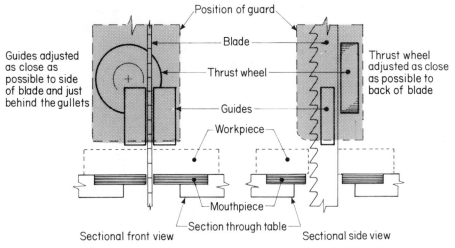

Position of guard

Blade

Thrust wheel

Guides

Workpiece

Guides adjusted
as close as
possible to side
of blade and just
behind the gullets

Thrust wheel
adjusted as close
as possible to
back of blade

Mouthpiece

Section through table

Sectional front view   Sectional side view

Note: Gap between guides and blade to be as close as practicable, but <u>not</u>
touching when the blade is running and not cutting

Fig. 5.17 Bandsaw - guard,
guide, thrust wheel and table

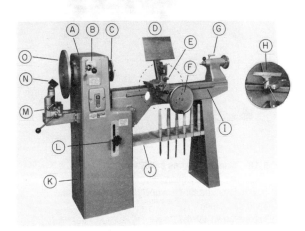

Fig. 5.18 Dominion 152 mm centre multi-speed alloy and
wood turning lathe
A - Headstock
B - Spindle brake
C - Inside face plate
D - Drawing stand
E - Compound travelling tool carriage
F - Handwheel; carriage movement
G - Tailstock
H - Standard tool rest, for hand-held chisels
I - Lathe bed
J - Shelf with turning tools
K - Access to headstock column
L - Speed control
M- Tool rest support
N - Tool rest
O - Outside face plate

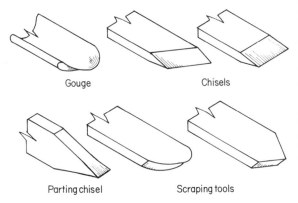

Gouge   Chisels

Parting chisel   Scraping tools

Fig. 5.19 Wood turning tools

soft alloy simply by turning the appropriate
handwheel.

For forming cylindrical shapes, the workpiece is
fixed between two centres – one in the headstock
which drives it round, and the other in the tailstock
which holds it steady. For bowl or disc shapes, the
workpiece is fixed to a face-plate.

Hand wood-turning tools have unmistakable
long handles which, when held under the forearm,
give good control over their use. These tools fall
into two groups. Those which have a true cutting
action include gouges, chisels (square and skew),
and parting tools. Those having a scraping
action – known as *scrapers* – have a flat face and
one under bevel and can be ground to whatever
shape is required, whether flat, V, or rounded.
Figure 5.19 shows typical blade shapes.

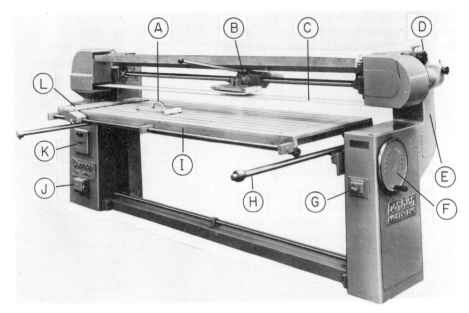

Fig. 5.20 Dominion under and over pad sanding machine

A - Hand pad
B - Travelling pressure pad
C - Sanding belt
D - Belt tracking and tensioning device
E - Swan-neck; accommodates long work
F - Handwheel; table rise and fall
G - Stop button
H - Table rails
I - Laminated wood table
J - Isolator
K - Control; start and stop buttons
L - Table fence

5.7 Sanding machines

The larger of these machines, like the one illustrated in Fig. 5.20, are used mainly to remove marks left by the rotary action of the planers and any ragged grain incurred during the planing process – also to dress (flatten) any uneven joints left after the assembly of such joinery items as doors and windows etc.

Smaller machines, like the combined belt and disc sander in Fig. 5.21, can be used with great accuracy for dressing small fitments, truing and trimming end grain, and sanding concave or convex surfaces.

A separate dust-extraction system is essential with all sanding machines.

Wood dust can not only be offensive but can also – depending on the wood species – cause dermatitis and become a contributory factor towards respiratory diseases.

5.8 Woodworking Machines Regulations 1974

These regulations impose statutory requirements on the manufacturers, owners, and users of woodworking machinery with regard to the provision and use of guards and safety devices. A suitable working environment must also be provided and maintained.

A full copy of the Woodworking Machine Regulations 1974 is printed within *Carpentry and Joinery 2*. However, a brief outline of the areas

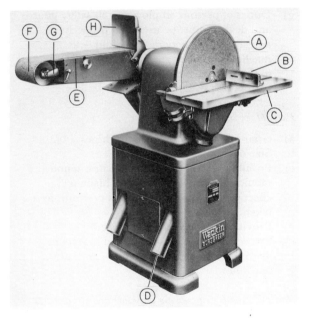

Fig. 5.21 Wadkin Bursgreen disc and belt sander

A - Sanding disc (405 mm dia.)
B - Sliding adjustable swivel fence
C - Sanding disc table (will tilt −10° to +45°)
D - Dust exhaust
E - Sanding belt table (horizontal or vertical positions)
F - Small diameter idler pulley (for sanding internal curves)
G - Sanding belt
H - Workpiece stop

covered by this sixteen-page document is as follows. Working operations of many of the machines are discussed in *Carpentry and Joinery 2*.

Part I: application, interpretation, and exemptions

a) Places where these regulations shall apply.
b) Types of machines covered by these regulations – which include portable (hand-held) machines.

Part II: all woodworking machines – general

a) Provision and construction of guards – their adjustment and maintenance.
b) Siting and position of machine controls.
c) Unobstructed working space.
d) Condition of workshop floors.
e) Temperature of rooms etc. where such machines are used.
f) Training of machine operators – no person will be allowed to operate a woodworking machine unless sufficiently trained to do so, as stated in Regulation 13.
g) Duties of persons employed – all safety guards and devices must be used and properly adjusted; any safety defect found – to machines, to their guards or safety devices, or to the surface of the machine-shop floor – must be reported to management.

Part III: circular-sawing machines

a) Guarding circular sawing machines.
b) Size of circular saw blades.
c) Prohibiting use for cutting rebates, tenons, moulds, or grooves unless the blade is effectively guarded and special precautions are taken.
d) Provision and use of push sticks.
e) Protection of persons employed to remove sawn material from the delivery ends of the machine.

Part IV: multiple rip-sawing machines and straight-line edging machines

Part V: narrow band-sawing machines

a) Provision of guards to saw wheels and blade.

Part VI: planing machines

a) Prohibiting use for cutting rebates, recesses, tenons, or moulds – unless the cutters are effectively guarded.
b) Types of cutter block.
c) Table gap – clearance between cutter and

surfacing tables.
d) Provision of bridge guards.
e) Adjustment of bridge guards for flatting and edging operations.
f) Providing cutter-block guards.
g) Providing and using push blocks.
h) Combined machines used for thicknessing.
i) Protection against ejected material.

Part VII: vertical-spindle moulding machines

Part VIII: extraction equipment and maintenance

a) Cleaning saw blades.
b) Provision for the extraction of chips and other particles.
c) Maintenance and fixing of permanent machines.

Part IX: lighting

a) Provision of adequate lighting – natural or artificial without glare.

Part X: noise

Provision of ear protection for persons employed at a machine for periods of eight hours or more at a noise level of 90 dB or more.

Fig. 5.22 Kango bench grinder
A - Specification plate
B - Motor
C - Eye shield
D - Grinding wheel
E - Adjustable tool rest
F - Base plate (fix to bench or stand)
G - Machine control
H - Grinding wheel guard

5.9 Grinding machines

Some kind of grinding machine is essential in the workshop, to carry out such functions as

a) re-forming the grinding angle on chisel and plane blades (see Figs 2.114 and 2.119),

b) regrinding and shaping hand and machine cutters,

c) resharpening screwdriver blades,

d) resharpening cold and plugging chisels.

All the above operations can be done on a dry grinding machine like the one in Fig. 5.22. Using this type of grinding machine requires considerable skill in preventing the part of the blade or tool which comes in contact with the high-speed abrasive wheel from becoming overheated (indicated by a blue colour) and loosing its temper (hardness). The end of the blade or tool should be kept cool by periodically dipping it into a dish of water, which should be close at hand.

Because of the small diameter of the grinding wheel, chisel and plane blades will have a grinding angle which is hollow (hollow ground), as seen in Fig. 5.23. This profile is preferred by many joiners as it tends to last longer than a flat-ground angle.

The problem of overheating can be overcome by using a *Viceroy* sharp-edge horizontal grinding machine, shown in Fig. 5.24. This machine has a built-in coolant system supplied with a special honing oil which flows continually over and through the surface of the grinding wheel while it is in motion. This machine is primarily used for grinding flat chisel and plane blades; however, there are attachments available which allow gouges and small machine blades to be ground.

The Abrasive Wheels Regulations 1970

These control safety in the use and installation of abrasive wheels, cylinders, discs, or cones. In so doing, they cover the following aspects;

a) maximum permissible speed of wheel to be specified – overspeeding could cause the wheel to burst;

b) proper mounting of the wheels;

c) appointment and training of persons to mount wheels;

d) provision and maintenance of guards and protection flanges;

e) effective means of starting and of cutting off the motive power;

f) workrests to be adjusted as close as practicable to the wheel whenever the machine is in use – otherwise the workpiece could become

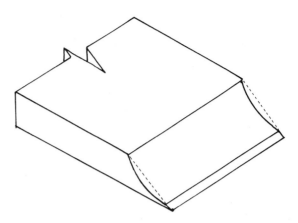

Fig. 5.23 Hollow ground blades

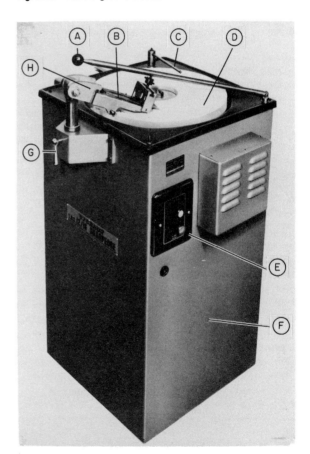

Fig. 5.24 Viceroy sharp edge machine
A - Operating lever
B - Plane lever holder; reverse side accommodates chisels
C - Honing oil distribution bar
D - Grinding wheel
E - Control; start and stop buttons
F - Cabinet; houses honing oil container and motor
G - Column height adjusts to suit angle of master arm
H - Master arm

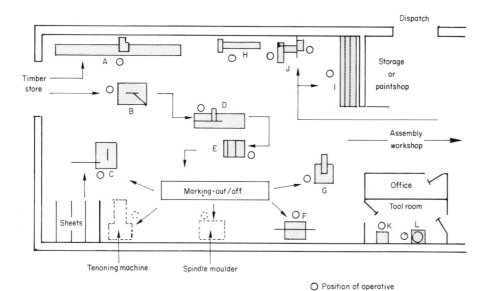

Fig. 5.25 Machine shop layout

A – Cross-cut saw
B – Circular-saw bench (rip and general purpose)
C – Dimension-saw bench
D – Hand-feed planer and surfacer
E – Roller-feed planer and thicknesser

F – Mortiser
G – Band-saw
H – Wood turning lathe
I – Under and over belt sander
J – Belt and disc sander

K – 'Dry' grinding machine
L – Horizontal 'wet' grinding machine

jammed between the wheel and rest, causing serious injury;

g) condition of the floor around where the machine is to be used.

This short eight-page document should be studied in full, together with the Protection of Eyes Regulations 1974, which stipulate among other things that persons who carry out dry grinding, wheel dressing (an operation which removes metal particles which have become embedded in the wheel), or truing (to keep the wheel concentric with the spindle) must wear approved eye protectors or be protected by a suitable transparent screen or shield against flying particles.

It is, however, advisable to be protected by both a screen and goggles, in case of an unsuspected ricochet.

5.10 Workshop layout

Before any decision is made with regard to workshop layout, some if not all of the following factors should be considered:

a) size of firm;
b) type of work;
c) available space (a clear space of 900 mm plus the maximum length of material to be handled

should be allowed around three sides of every machine);
d) work-force – whether full-time wood machinists are to be employed;
e) number and type of machines likely to be cost effective;
f) methods of providing chip- and dust-extraction systems;
g) provision for a tool room (for tool and machine maintenance);
h) storage and racking facilities;
i) suitable and adequate lighting;
j) suitable and sufficient power supply.

Ideally, for a machine shop to work with maximum efficiency, machines should not occupy valuable space unless they are used regularly or contribute towards a steady flow of jobs through the workshop.

Figure 5.25 shows how the machines mentioned in this chapter could be positioned to produce a work flow to suit a small to medium-sized joiners shop. The overall layout does, however, allow for the inclusion of a tenoner and a spindle moulder at a later date. These extra machines (dealt with in *Carpentry and Joinery 2*) could be regarded as being essential if the machine shop were to produce joinery items on a production basis.

6

Basic woodworking joints

There are many different joints that the carpenter and joiner may use. This chapter is concerned mainly with those made by hand.

Joints generally fall into three categories and carry out the following functions:

	Category	Joint	Function
a)	Lengthening	End	To increase the effective length of timber
b)	Widening	Edge	To increase the width of wood or manufactured boards
c)	Framing	Angle	To terminate or to change direction

6.1 Lengthening - end joints (Fig. 6.1)

Where timber is not long enough, a suitable end joint must be made.

Butt joint   Figure 6.1(a) shows an end butt joint with cleat. This method would be used only where one face could be concealed.

Scarf joint   Figure 6.1(b) shows two methods of making a scarf joint. For structural use they will require a slope of 1 in 12 or less. The second method incorporates a hook which enables the joint to be tightened with folding wedges.

Laminated joint   By laminating different lengths of timber together with nails and/or glue, large long lengths of timber can be manufactured. Basic principles are shown in Fig. 6.1(c).

Finger joint   A finger joint is shown in Fig. 6.1(d). This is produced by machine, then glued and assembled by controlled end pressure. This is a useful method of using up short ends and upgrading timber – after the degraded portion or portions have been removed, the remaining pieces are rejoined.

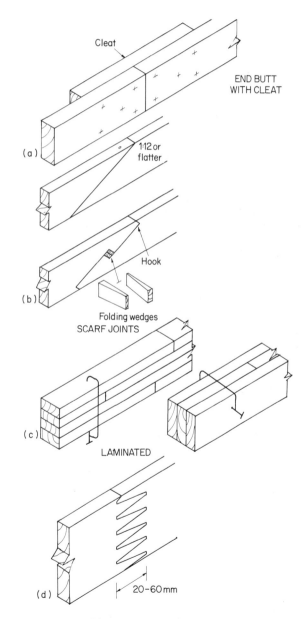

FINGER JOINT (formed by machine)

Fig. 6.1 Lengthening joints (see also half lap and sloping halving in Fig. 5.11)

Half-lap joint   See also the half-lap joints featured in Fig. 6.11.

6.2 Widening – edge joints (Fig. 6.2)

Examples of these joints are shown in Fig. 6.2. They all enable a board's width to be increased, but whether the joint is to be permanent (glued) or flexible (dry joint) will depend on its location.

Whatever method of joining is chosen, it is always wise to try to visualise how a board will react if subjected to moisture. As stated in Chapter 1, tangential-sawn boards are liable to 'cup'. Figure 6.3 shows a way to minimise this effect.

Wide solid wood boards cannot generally cope with situations such as floors, walls, ceilings, or doors, etc. as their moisture content is liable to become unstable; therefore a flexible method is

used like those shown in Fig. 6.4, thus reducing the risk of splitting.

Butt joint   This is the simplest of all edge joints and is the basis of all the other forms shown in Fig. 6.2. If the joint is to be glued, it is important that the adjoining edges match perfectly. Figure 6.5 shows how this is achieved. The boards are first marked in pairs (Fig. 6.5(a)), then each pair is planed straight and square by using a long-soled try-plane (Fig. 6.5(b)). They are then repositioned edge to edge to check that no light shows through the joint and both faces are in line (Fig. 6.5(c)).

Glue is applied while both edges are positioned as if they were hinged open. They are then turned edge to edge and rubbed one on the other to remove any surplus glue and finally form a bond – hence the term *rubbed joint*.

Dowelled joint   By inserting dowels at approximately 300 mm intervals, the butt joint can be both strengthened and stiffened. Figure 6.6 shows how the dowels are positioned and how provision is made for the escape of glue which may become trapped in the hole. (See also Fig. 6.17.)

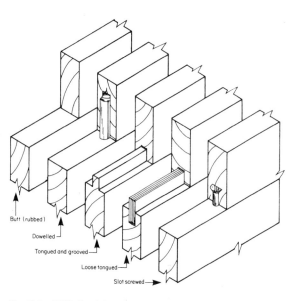

Fig. 6.2 Widening joints

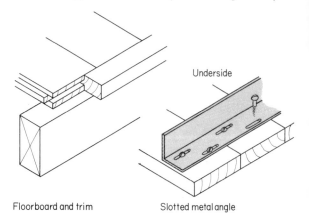

Floorboard and trim   Slotted metal angle

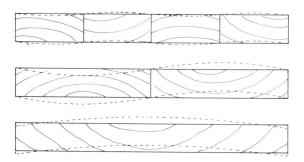

Fig. 6.3 Balancing the effect of moisture movement

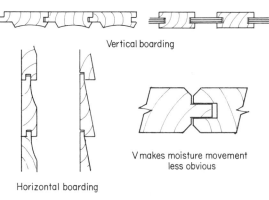

Fig. 6.4 Joints which are allowed to move

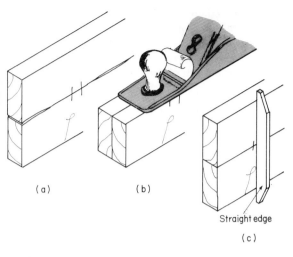

(a) (b)

Straight edge

(c)

Fig. 6.5 Planing a butt (rubbed) joint

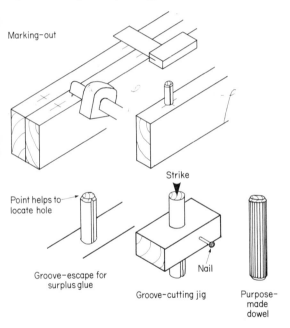

Marking-out

Strike

Point helps to
locate hole

Groove–escape for
surplus glue

Groove-cutting jig

Nail

Purpose-
made
dowel

Fig. 6.6 Preparing a dowel joint

Tongued-and-grooved joints – loose tongue
Another method of strengthening a butt joint is to
increase the surface area to be glued (the glue line).
These joints do just that. Cutters used to form
these joints are shown in Fig. 6.7.

Slot-screwed joint This is a simple yet effective
method of edge jointing which can also be used
whenever a secret fixing is to be made. Figure 6.8
shows how the joint is made. One board is offset by
the amount that the screw will travel in the slot
(10–20 mm, depending on the screw length and

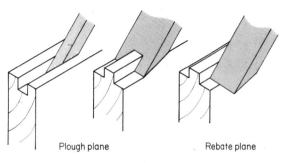

Plough plane Rebate plane

Fig. 6.7 Forming a tongue and groove

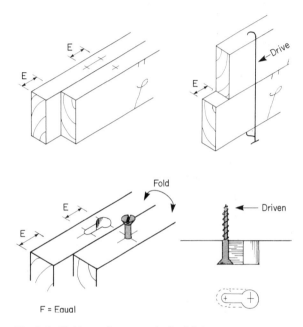

E E ←Drive

E E

Fold

E ←Driven

E

F = Equal

Fig. 6.8 Making a slot-screwed edge joint

gauge) and, when one board is driven over the
other, the screw head bites into the slot and
becomes firmly embedded.

6.3 Framing – angle joints

Joints used to form angles and/or junctions can be
divided into the following groups:

a) housing,
b) halving,
c) mortise and tenon,
d) bridle,
e) dowelled,
f) notched and cogged,
g) dovetail,
h) mitre and scribe.

Housing joints The simplest and probably most common housing joint is the *through* housing (Fig. 6.9(a)), which gains bearing support from the notch when formed vertically, and resists side movement when used horizontally as shown. *Stopped* housings (Fig. 6.9(b)) conceal the trench on one edge, and *double-stopped* housings (Fig. 6.9(c)) conceal it on both. *Dovetailed* housings (Fig. 6.9(d)) have one or both (not illustrated) sides of the trench sloping inwards, thereby adding part or total resistance to withdrawal.

With the exception of that in Fig. 6.9(d), these joints generally require nailing.

Figure 6.10 illustrates the steps in forming the trench for a stopped housing. They are;

1 mark and gauge the width and depth,
2 bore two or more holes to the width and depth of the trench,

3 chop the edges of the holes square,
4 using the 'toe' of a tenon saw, make two or more saw kerfs to the depth line,
5 remove waste wood with a chisel and mallet,
6 level the bottom of the trench with a router.

Housing joints are often used in the construction of shelf and cabinet units, partitions, and sectional timber-framed buildings.

Halving joints These are used where timber members are required either to cross or to lap each other. Named examples, most of which are exploded, are shown in Fig. 6.11, where it will be seen that each cut-away portion corresponds with the part it is to join. It is worth noting that the 'half lap' can be used either at a corner – as shown – or as an end joint (end lap) similar to the *sloping halving* although not as strong.

The joint is made by a combination of those methods used in forming the housing, tenon, and dovetail joints.

Mortise-and-tenon joints These are the most common of all conventional framing joints –

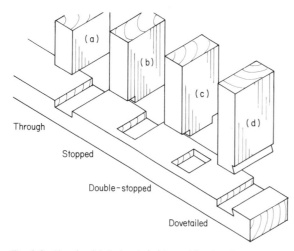

Fig. 6.9 Housing joints (exploded isometric views)

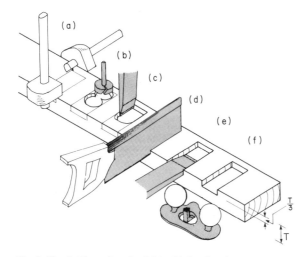

Fig. 6.10 Cutting a housing joint with hand tools

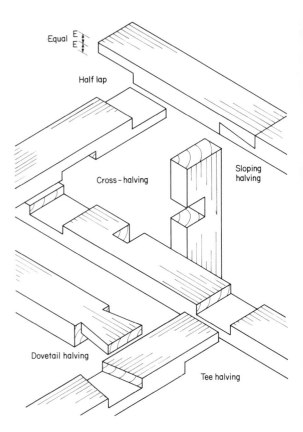

Fig. 6.11 Halving joints (exploded isometric views)

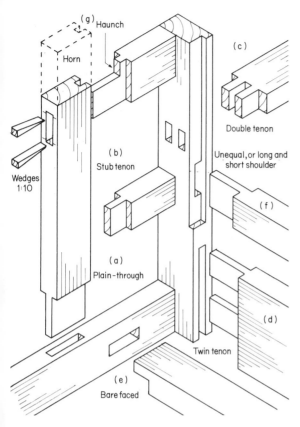

Fig. 6.12 Mortise-and-tenon joints (exploded isometric views)

probably due to their versatility and easy concealment. Figure 6.12 illustrates in exploded detail a few examples of how and where these joints are formed and used. The joint's name usually reflects its size, shape, or position. For example;

a)	*through* mortise and tenon	the mortise hole goes completely through the material;
b)	*stub* tenon	mortised only part way into the material;
c)	*double* tenon	two tenons side by side (usually in the thickness of a member);
d)	*twin* tenon	two tenons cut in the depth of a member;
e)	*bare-faced* tenon	only one shoulder;
f)	*unequal shoulder*	one shoulder longer than the other.

(Note: *double* tenon may be referred to as *twin* tenon and vice versa.)

Haunches (Figs 6.12(g) and 6.13)   These serve two purposes; they prevent joints becoming *bridles* (Fig. 6.16) and they reduce the length of the

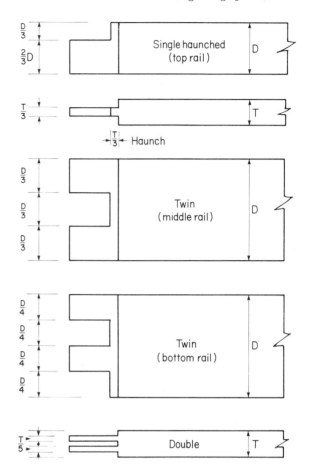

Fig. 6.13 Proportioning a tenon

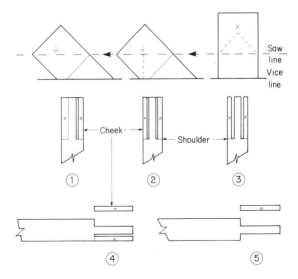

Fig. 6.14 Sequence of cutting a tenon or half lap

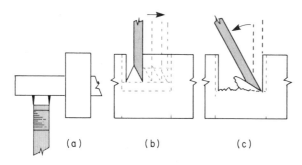

(a)   (b)   (c)

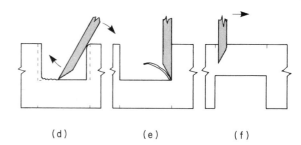

(d)   (e)   (f)

Fig. 6.15 Chopping a mortise hole (sectional details) - back lines of mortise holes omitted for the sake of clarity

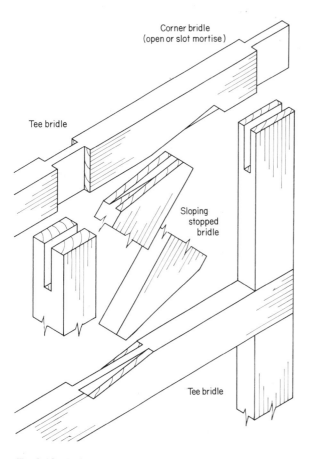

Corner bridle
(open or slot mortise)

Tee bridle

Sloping
stopped
bridle

Tee bridle

Fig. 6.16 Bridle joints (exploded isometric views)

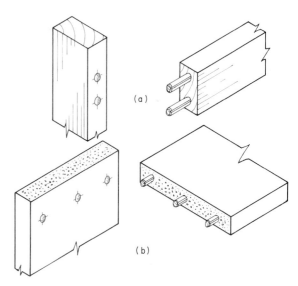

(a)

(b)

Fig. 6.17 Dowelled joints

Figure 6.15 illustrates a sequence of operations for chopping out a mortise hole:

1   Set the mortise gauge to the width of the chisel, which should be as near as possible to one third the width of the material being mortised.
2   After having secured the workpiece, chop a hole approximately 15 to 20 mm deep and 4 mm in from one side.
3   Working from left to right or from right to left, chop and gently lever waste wood into the hole formed.
4   Repeat until midway into the workpiece, carefully lifting out chippings at each level. (Note how the waste left on the ends of the hole prevents damage to the mortise hole during this process).
5   Square the ends of the mortise hole.
6   Turn the workpiece over and repeat the process. (The bench should be protected at this stage if a through mortise is to be cut.) See also Fig. 2.81.

mortise hole which would otherwise be required for wide rails. Figure 6.13 is a general guide as to how tenons and haunches are proportioned.

Figure 6.14 shows a method of cutting a tenon. The same principles apply to the *half lap* and *bridle*, although in the latter case, where there are no cheeks, care must be taken to cut on the waste side of the lines. (See also Figs 6.17 and 6.19.)

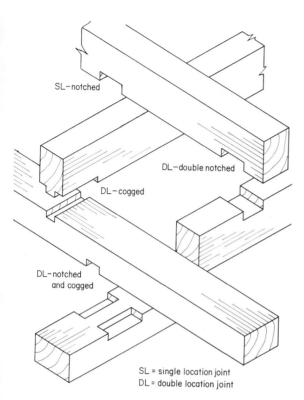

Fig. 6.18 Notched and cogged joints (exploded isometric views)

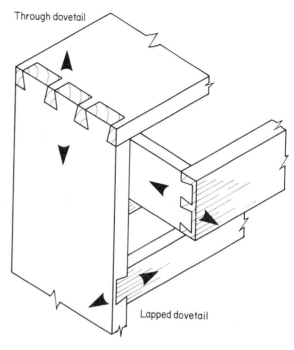

Fig. 6.19 Dovetail joints

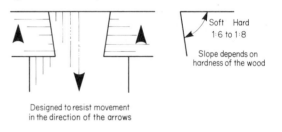

Designed to resist movement in the direction of the arrows

Bridle joints Except for the 'corner' bridle – also known as an *open* or *slot* mortise – bridle joints slot over through-running members. Named examples are shown in Fig. 6.16. They are cut in a similar manner to tenon and halving joints.

Dowelled joints These are useful alternatives to mortise-and-tenon joints for joining members in their thickness (Fig. 6.17(a)) or as a means of framing members in their width (Fig. 6.17(b)). Dowel and hole preparation is similar to the methods shown in Fig. 6.6. Correct alignment of dowel with hole is critical, but this problem can be overcome by using a dowelling jig (a template and guide for boring holes accurately).

Notched and cogged joints As shown in Fig. 6.18, notches are used to locate members in one or both directions and as a means of making any necessary depth adjustments (joist to wallplates etc.). Cogged joints perform a similar function, but less wood is removed, therefore generally leaving a stronger joint. They do, however, take much longer to make.

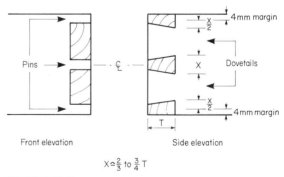

N.B. GUIDE ONLY

Fig. 6.20 Proportioning a dovetail joint

Dovetail joints Figure 6.19 shows how dovetailing has been used to prevent members from being pulled apart. The strength of a dovetail joint relies on the self-tightening effect of the

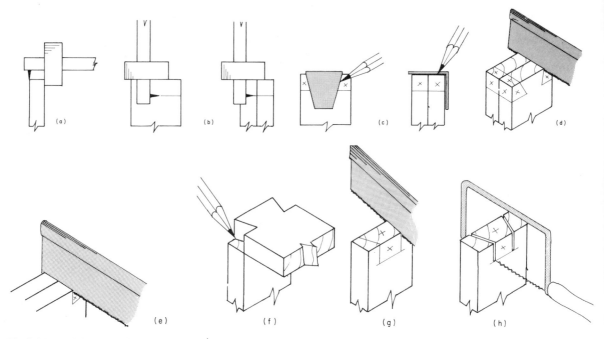

Fig. 6.21   Marking-out and cutting a dovetail joint

dovetail against the *pins*, as shown by the direction of the arrows.

Dovetail slope can vary between 1 in 6 to 1 in 8, depending on the physical hardness of the wood. The number of dovetails and their size will vary with the width of the board. The dovetails are usually larger than the pins (except those produced by machine – which are of equal width), and a guide to their proportions is given in Fig. 6.20.

Figure 6.21 shows a method of marking and cutting a single through dovetail;

a)   Set the marking gauge to the material thickness.

b)   Temporarily pin together those sides (in pairs) which are to be dove tailed at their ends. Gauge all round.

c)   Using a bevel or dovetail template, mark the dovetails (see Fig. 2.12).

d)   Cut down the waste side to the shoulder line, using a dovetail saw.

e)   Cut along the shoulder line and remove the cheeks.

f)   Divide the sides and mark off the pins.

g)   Cut down to the shoulder line with a dovetail saw.

h)   Remove waste with a coping saw, then pare square with a chisel.

The joint should fit together without any further adjustments!

Some joiners prefer to cut the pins first and the

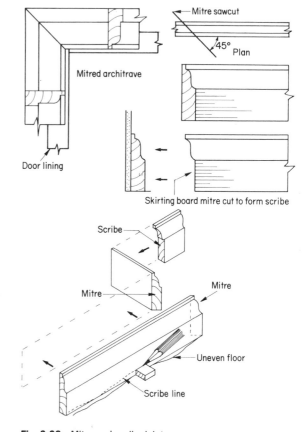

Fig. 6.22   Mitre and scribe joints

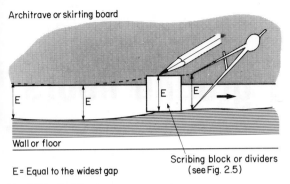

Architrave or skirting board

E   E   E   E   →

Wall or floor

E = Equal to the widest gap

Scribing block or dividers
(see Fig. 2.5)

Fig. 6.23 Scribing to an uneven surface

dovetails last. Both methods are acceptable, but the sequence described is quicker and tends to be more accurate. Probably the first opportunity the student gets to use this joint is during the construction of a tool box, chest or case – dovetail joints should be the first choice when constructing one or more units of the *Porterbox* projects.

Mitre and scribe joints   These are generally associated with joining trims – i.e. cover laths, architraves, skirting boards, and beads – either at external or internal angles. The joint allows the shaped sections to continue round or into a corner, as shown in Fig. 6.22.

A mitre is formed by bisecting the angle formed by two intersecting members and making two complementary cuts. A scribe joint has its abutting end shaped to its own section profile, brought about by first cutting a mitre (Fig. 6.22). (See also *Carpentry and Joinery 2* and *3* for hand and machine scribes.)

Scribing   Where a joint has to be made against an uneven surface, such as a floor, wall, or ceiling, scribing provides a means of closing any gaps. Figures 6.22 and 6.23 show that, by running a gauge line parallel to the uneven surface, an identical contour will be produced (see also Fig. 6.5).

7

Suspended timber ground floors

As can be seen from Fig. 7.1 this type of floor – also known as a 'hollow' floor – is made up of a series of timber beams, called *joists*, covered with boards. The whole floor is then supported by wallplates resting on purpose-built *honeycombed* sleeper walls (Figs 7.1 and 7.2). Alternatively, the joist ends can be built into the inner leaf of the perimeter cavity walls (Fig. 7.3), provided they are protected from cavity moisture. Note: thermal insulation is not shown – see *Carpentry and Joinery 2* for details.

These floors have of recent years been the subject of many outbreaks of dry rot (see *Carpentry and Joinery 2*), primarily due to the timber used in their construction having been allowed to come into contact with sufficient amounts of moisture to raise its moisture content (m.c.) above the danger level of 20%. This problem can often be traced to the omission or breakdown of the DPC (damp-proof course – a continuous layer of thin impervious material sandwiched between the brick or blockwork) which acts as a barrier against any

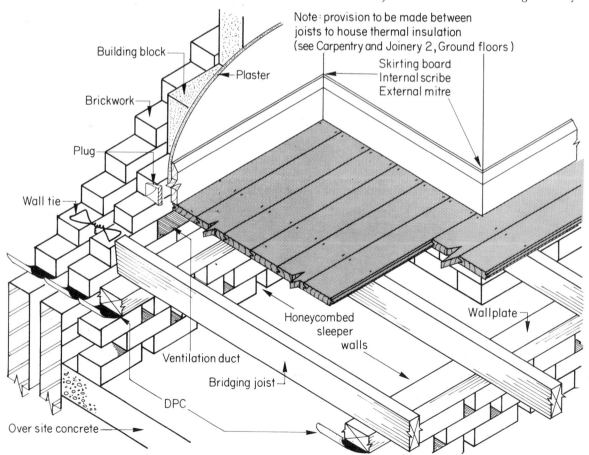

Building block

Plaster

Brickwork

Plug

Wall tie

Note: provision to be made between joists to house thermal insulation (see Carpentry and Joinery 2, Ground floors)

Skirting board
Internal scribe
External mitre

Honeycombed
sleeper
walls

Wallplate

Ventilation duct

Bridging joist

DPC

Over site concrete

Fig. 7.1 Suspended timber floor

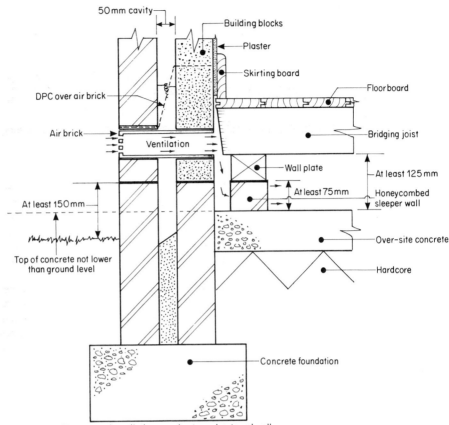

Note: actual ventilation opening to each external wall
at least 3000 mm² for each metre run of wall

Fig. 7.2 Using sleeper walls
to support floor joists

rising damp. Alternatively, it may be due to
moisture-laden air having been allowed to
condense on the underside of the floor, due to
inadequate ventilation of the subfloor space. Free
circulation of air to the whole of the subfloor space
is therefore essential, and is achieved by the
provision of air bricks (purpose-made perforated
blocks, Fig. 7.2) in the outer perimeter walls and
the honeycombing of sleeper walls. Dividing walls
must also be strategically pierced.

Construction of these floors is strictly controlled
by the Building Regulations 1985 to ensure both
structural stability and protection against the
ingress of moisture (Fig. 7.2).

To achieve the correct overall balance in the
design of the floor, the following factors covered by
approved document A must be taken into account;

1   clear span of joist,
2   joist sectional size,
3   joist spacing,

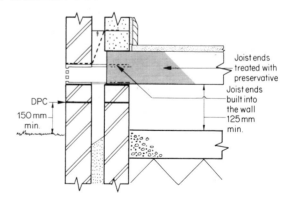

Fig. 7.3 Building-in floor joists (not recommended)

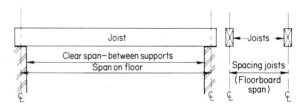

Fig. 7.4 Floor joist - span and spacing

4   grade of timber to satisfy strength class (SC),
5   dead load supported by the joist,
6   flooring material,
7   flooring-material thickness.

7.1 Floor joists

The floor design is usually such that the joist is laid to bridge the shortest distance between room walls. This distance may however be further reduced by the introduction of sleeper walls. The effective span of a bridging joist is shown in Fig. 7.4.

A joist end section in relation to its span must be strong enough not only to withstand the dead weight of the floor but also to safely support any load that may be placed upon it. The importance of stating timber sizes on a drawing as *length ×* *width × depth* (or *thickness*) should be stressed at this point, particularly for those sections which are expected to withstand heavy loads. Figure 7.5 shows the effect of dimensioning out of sequence.

Joist spacing means the distance between the centres of adjacent joists, as shown in Fig. 7.4. These centres range between 400 and 600 mm, depending on the factors previously stated. However, joist centres should start and finish 75 mm in from the wall face (providing the joists are 50 mm wide). This will leave a 50 mm gap between the wall and the joist running parallel to it, so allowing air to circulate freely.

Levelling

The levelling process can be carried out by one or a combination of the following methods;

a)   spirit-level and straight-edge,
b)   water (aqua) level (see *Carpentry and Joinery 2*),
c)   optical levelling (see *Carpentry and Joinery 2*).

Figure 7.6 shows four simple stages that can be adopted with a spirit-level and straight-edge, provided there is sufficient room to work. If, however, restrictions prevent the joists from being sighted over, then use a spirit-level at stage 3 and a straight-edge to line through at stage 4. (A straight

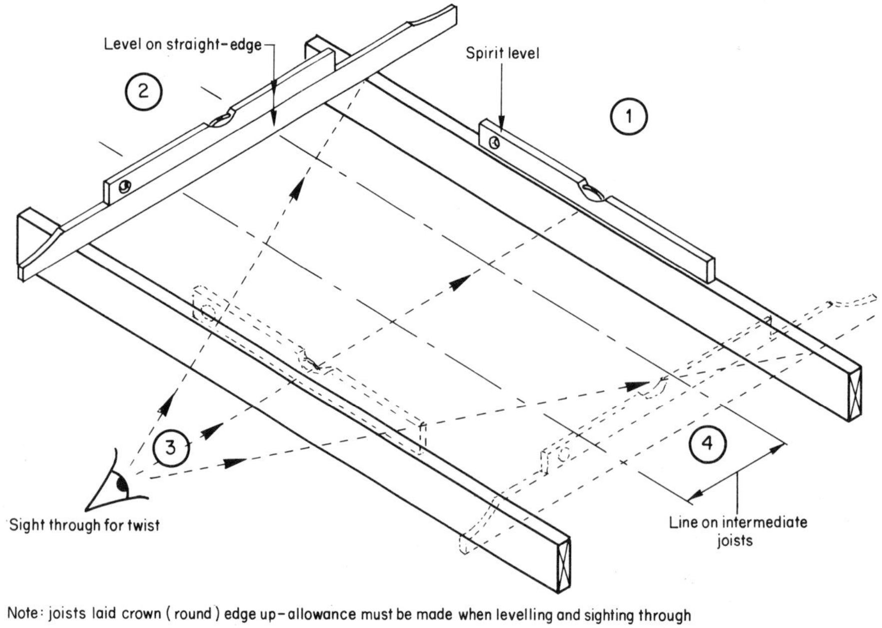

2000 x 150 x 50

Example size only

2000 x 50 x 150

Note: the order of dimensioning drawings:-length x width x depth

Fig. 7.5 Floor joist end section in relation to its span

Fig. 7.6 Laying and levelling floor joists

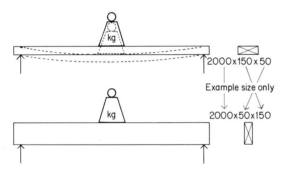

Level on straight-edge

Spirit level

② ①

③ ④

Sight through for twist

Line on intermediate joists

Note: joists laid crown (round) edge up – allowance must be made when levelling and sighting through

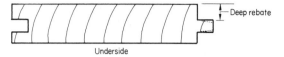

Fig. 7.8 Floor board section

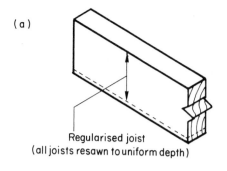

(a)

Regularised joist
(all joists resawn to uniform depth)

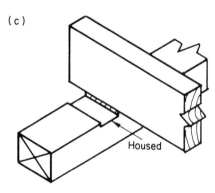

(b)

Packed

Packing must
be well nailed
to wallplate

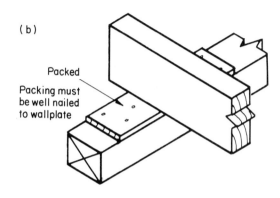

(c)

Housed

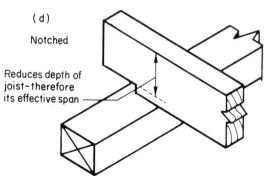

(d)

Notched

Reduces depth of
joist—therefore
its effective span

Fig. 7.7 Levelling and bearing adjustments

length of floor board will act as a suitable straight-edge.)

The accuracy of the whole process will depend very much on whether the wallplates (if used) were levelled correctly in the first instance. If the wallplates are level and the joists are all the same depth (regularised, Fig. 7.7(a)) the bearing of the joists should only require minor adjustment.

If, however, it is found that adjustments must be made, Figs 7.7(b) and 7.7(c) show acceptable methods. It should be noted that the method shown in Fig. 7.7(d) will reduce the depth of the joist and therefore its efficiency over that span (see Fig. 7.4).

The use of wallplates not only enables the floor's weight to be more evenly distributed over a wider area but also provides a means of securing the joists by nailing.

If the joist ends are to be built into the walls (Fig. 7.3) it is advisable to use slate not wood as a packing, because the wood packing could shrink and eventually work loose. Alternatively, the method shown in Fig. 7.7(d) may be adopted, provided the joist is deep enough to allow this method to be used without it being weakened. However, when the joists are level and correctly spaced, they must be kept that way *by tacking a spacing lath on to the tops of the joists*. This is removed after the joists have been walled in by the bricklayer.

7.2 Flooring (decking)

The flooring material will consist of either;

a) planed, tongued, and grooved (p.t.g.) floor board,

b) flooring-grade chipboard – (see *Carpentry and Joinery 2*),

c) flooring-grade plywood (see *Carpentry and Joinery 2*).

This volume deals with floor boards, whereas sheet flooring materials will be dealt with in *Carpentry and Joinery 2*.

Figure 7.8 shows a typical section of p.t.g. softwood board, with a deeper rebate cut from the

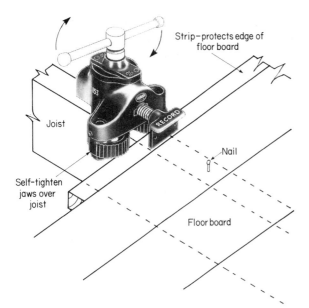

Fig. 7.9 Record flooring cramp and its use

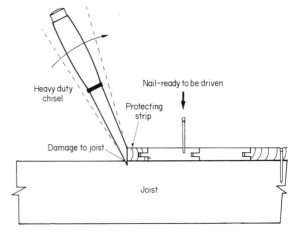

Fig. 7.10 Closing floor boards by leverage

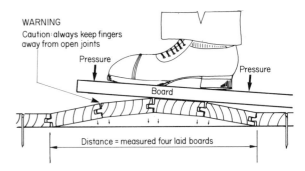

Fig. 7.11 Folding method of laying floor boards

upper face than from the underside. This deep rebate provides the joiner with a quick and useful guide to the board's face side. Also, before overall floor coverings (carpets etc.) became fashionable and less expensive, it was commonplace to walk directly on the floor boards – the only protection offered being a layer of varnish and/or wax polish. The story has it that the deeper rebate therefore allowed for more wear to take place before the tongue and grooves became exposed and the boards dangerously thin.

Laying floor boards

Boards are laid at right angles to the joists, and wherever they cross a joist they are eventually double-nailed to it – using either lost-head nails or flooring brads (Section 15.1). Nails should be $2\frac{1}{2}$ times the thickness of the floor board, whereas flooring brads are of standard cut length.

The first board should be nailed approximately 15 mm away from one wall (the gap later being covered by skirting), then three to six boards are positioned edge to edge against it – the number of boards will depend on the method of clamping. Figure 7.9 shows a typical flooring clamp (cramp) and its application. In the past, lever and folding methods have been used (Figs 7.10 and 7.11). Although not as efficient as the flooring clamps, these methods are still effective and are useful when dealing with small areas or working in confined spaces. However, the chisel does damage

the top edge of the joist. Once the boards have been cramped tight (not overtightened), they should be nailed to the joist or be spot nailed sufficiently to hold the boards in position.

Where the boards have to be end jointed, a splayed end joint can be used (Fig. 7.12(a)). This helps to prevent splitting. These joints should always be staggered as shown in Fig. 7.12(b) – not as Fig. 7.12(c) – unless a trap has to be left in the floor for access to the subfloor space (Fig. 7.13), in which case it should be sited in a non-traffic area, for example a cupboard under the stairs etc.

Continue in this way across the room, where the last few boards can be cramped using the methods shown in Figs 7.10 and 7.11. The spot nailing should have been sufficient to indicate the centre position of *all* the joists, and this now enables the joiner to mark a series of pencil or chalk marks across the room as a guide for the final nailing-down process, followed by punching the nail heads just below the surface. (The final nailing-down process is often termed *bumping*.) Using the spot-nailing method releases the flooring clamps for use in other rooms.

Note: if service pipes have been laid in close proximity to the floor joists, there is a danger of driving nails into them. It would therefore be advisable to nail that area fully as the floor boards are being laid, rather than trying to remember where the pipes were or forgetting to mark the danger area.

7.3 Skirting

After the plastering is complete, a set of new operations takes place – known as *second fixing*. One of these operations is to fix the skirting-board.

Skirting-boards are machined to many different profiles, some of which are illustrated in Fig. 7.14. The profile usually reflects the class of work being carried out.

The main function of a skirting-board is to provide a finish between the wall and floor. It also acts as a seal against draughts from the subfloor

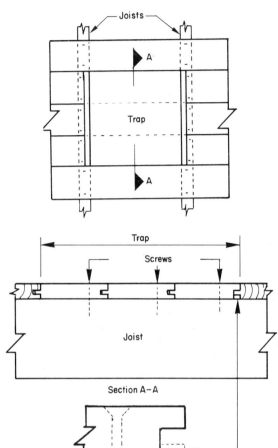

Fig. 7.13 Forming access and service traps

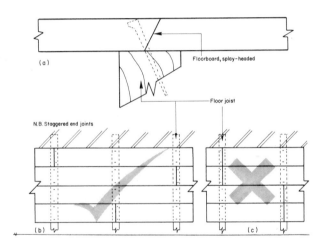

Fig. 7.12 End jointing of floor boards

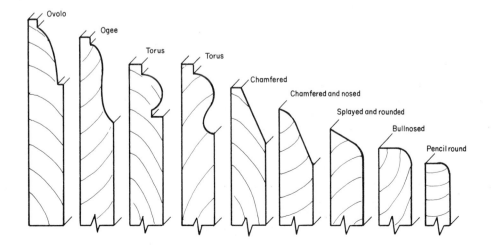

Fig. 7.14 Skirting board sections

space and as a buffer to protect the wall covering from knocks that may occur during cleaning the floor covering – timber being much more resilient than plaster.

Fixing

Although skirting board is only a 'trim' (cover), because of its frequent ill-treatment it does require to be fixed firmly to the walls. This is usually achieved by one of three ways (see also Chapter 15);

a) plugging (Fig. 7.15(a)),
b) timber grounds (Figs 7.15(b) and 7.15(c)),
c) direct nailing.

The use of wood plugs in brickwork is a very useful method of fixing to walls where brick or dense concrete blocks have been used. (Figures 15.10 and 15.11 show the method of preparing and fixing such plugs.)

Plastics plugs should also be considered as a modern alternative where fixing by screws or screw nails is practicable or a permitted alternative (see Chapter 15).

Figures 7.15(b) and 7.15(c) show the use of timber grounds as a fixing medium. The method of Fig. 7.15(b) provides a finish for the plasterwork and a continuous longitudinal fixing, while leaving a narrow service duct behind the skirting; whereas that in Fig. 7.15(c) is employed when used in conjunction with wall panelling.

Certain modern types of building blocks will permit cut clasp nails to be driven directly into them, offering sufficient holding power for light-sectioned skirting-board, thus speeding up the whole fixing process.

Joints

There are only three joints to consider here;

1 butt joint,
2 mitre joint (Fig. 6.22),
3 scribed joint (Fig. 6.22).

The layout of the floor will dictate the position and type of joint employed. For example, at doorway openings it is common practice to butt the skirting board up to the door-casing trim (known as the architrave); external corners are mitred; and internal corners are scribed (Figs 7.1 and 6.22). However, where a piece of skirting-board has to be joined in its length (a practice to be discouraged – it is not usually necessary), a cut of 45° is more desirable than a 90° butt joint. A cut of 45° enables the joint to be nailed more efficiently and provides a partial mask if or when shrinkage occurs (Section 1.7).

The methods of cutting these joints is dealt with in Section 6.3, together with the technique for scribing timber along its length – necessary where a skirting-board has to be fitted to an uneven floor surface (Figs 6.22 and 6.23).

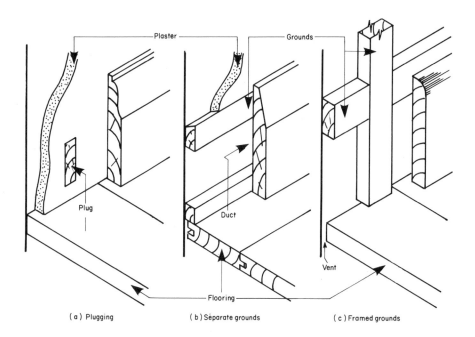

Fig. 7.15 Methods of fixing skirting boards to walls

(a) Plugging   (b) Separate grounds   (c) Framed grounds

8

Gable-ended Traditional single roofs

The roof must be included in the design of a building as a whole, be it a factory, a house, or a garage. Just as its substructure depends on the roof for cover, so must the roof rely on the walls for support. Therefore the roof and walls must complement each other. Size, shape, location, use, and appearance are all factors which influence the choice of roof.

This chapter is concerned with roofs which span up to 4 metres, have surfaces steeper than 10° to the horizontal – known as *pitched* roofs – and have their length terminated by end walls (gable walls).

NB: Trussed rafter roofs are dealt with in Book 2.

8.1 Roof terminology

Figure 8.1(b) names those parts (elements) of a pitched roof to which the various members (components) shown in Fig. 8.1(c) relate.

Single roof – common rafters span from wallplate to ridge board without intermediate support.

Double roof – common rafters have their effective span halved by mid support of a beam (purlin). Also known as a *purlin* roof (discussed in *Carpentry and Joinery 2*).

Roof span – usually taken as the distance between the outer edges of the wallplates (Fig. 8.1(a)).

Roof rise – the vertical distance from a line level with the upper surfaces of the wallplates to the intersection of the inclined slopes (Fig. 8.1(a)).

Roof pitch – the slope of the roof (Fig. 8.1(a)) expressed either in degrees or as the fraction rise/span, i.e. rise divided by span. For example, for a rise of 1 m and a span of 2 m,

$$\text{Roof pitch} = \frac{\text{Rise (1 m)}}{\text{Span (2 m)}} = \frac{1}{2} \text{ pitch or } 45°.$$

The roof covering, i.e. tiles, slates, etc., will determine the pitch to be used.

Roof elements (Fig. 8.1(b))

Ridge (apex) – the uppermost part of the roof.
Gable – the triangular upper part of the end walls.
Verge – the overhanging edge at the gable ends.
Eaves – the area about the lower edge of the roof surface at the top of the outer face walls.

Roof components (Fig. 8.1(c))

Ridge board – receives the ends of the rafters.
Common rafter (spar) – rafter spanning from wallplate to ridge board.
Gable ladder – framework taking the roof over the gable wall.
Ceiling joists – carry ceiling material and act as rafter ties.
Wallplate – provides a fixing for rafters and ceiling joists and distributes the roof load.
Barge (verge) board – cover (trim) for the ends of horizontal members, i.e. wallplate, gable ladder, and ridge board.
Fascia board – vertical facing to rafter ends (spar feet).
Soffit board – horizontal board closing the underside of the spar feet.
Gutter – channel section in metal, plastics, or wood fixed slightly off level to the spar feet to allow roof water to drain away to a fall pipe (downpipe). Wooden gutters are sometimes called *eaves spouting*.

8.2 Forming a pitched roof

If a length of timber were reared against a wall, like a rafter in a lean-to roof (Fig. 8.4(b)), or leant

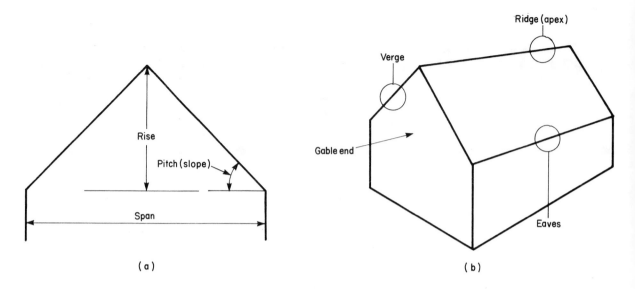

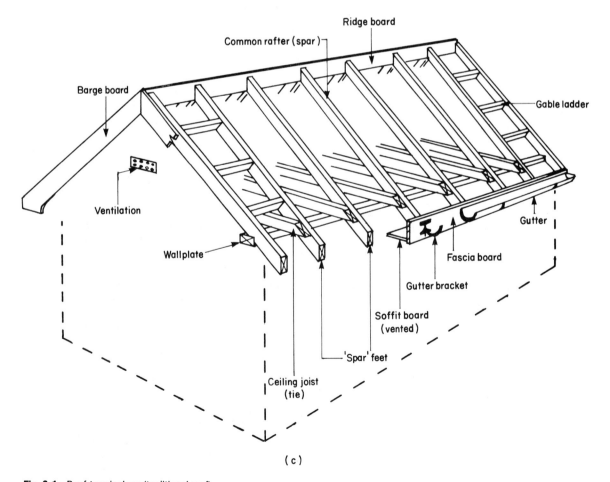

Fig. 8.1   Roof terminology (traditional roof)

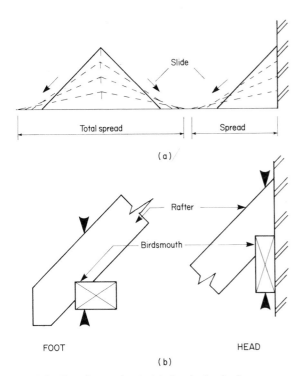

Slide

Total spread

Spread

(a)

Rafter

Birdsmouth

FOOT

HEAD

(b)

Fig. 8.2 The effect and restraint of an inclined rafter

against a similar inclined length to form the ridged effect of a couple roof (Fig. 8.4(c)), it would become apparent that the slope, length, and weight, together with the smoothness of the surfaces in contact, would lead to the sliding effect shown in Fig. 8.2(a).

Provided the rafters have bearings at the top and bottom of the roof which are capable of transmitting loads vertically, like those offered by monopitch and lean-to roofs (Figs 8.4(a) and (b)), slide can be restrained by cutting a *birdsmouth* into the rafter at these points – as shown in Fig. 8.2(b). However, deflection (due to loading) of the rafters and the hinging effect at the apex of a couple roof could produce enough horizontal thrust to push the supporting walls outwards; therefore the walls must be strengthened or a tie be introduced into the roof structure.

These principles can be more fully understood by using a model similar to that shown in Fig. 8.3(a). This consists of four narrow strips of plywood – two rafters, one tie, and a strut – with holes bored to suit either dowel or nuts and bolts (free to move). The walls are represented by short wide strips.

The arrangement can be set up as follows:

1 Fig. 8.3(b) – when the pitch is varied,

instability is very noticeable as the walls move away.
2 Fig. 8.3(c) – the introduction of a high collar (tie) produces severe bending of the rafters at this point. The walls still move outward.
3 Fig. 8.3(d) – when the collar (tie) is lowered, the roof slope becomes more stable (Fig. 8.4(d)).
4 Fig. 8.3(e) – wall movement is prevented by fully lowering the tie (Fig. 8.4(e)).

Figure 8.4 identifies common methods of forming pitched roofs and offers a guide to their maximum span. Size of members will depend on pitch, span, and form. Where a tie acting as a ceiling joist (couple close roof) exceeds 2.5 m, it will, for reasons of the most economical sectional size (see Fig. 7.5), need to be restrained from sagging by using hangers and binders (Fig. 8.4(f)); otherwise the sag will put undue stress on the joints and unbalance the framework.

8.3 Common rafters

Figure 8.5 shows a method of obtaining the bevels for common rafter *plumb cut* and *seat (foot) cut*.

A triangle is drawn to a suitable scale (the larger the better) to represent the roof's rise and half its span (to the outer edges of the wallplates) – Fig. 8.5(a). The two angles formed by the common rafter are measured or transferred to a short end of timber by laying it over the drawing. The angles PC (plumb cut for common rafter) and SC (seat cut for common rafter) are clearly marked as shown in Fig. 8.5(b). Rafter length is also determined from the drawing and can be checked by using Pythagoras' theorem.

Using a sliding bevel, the angles are transferred to a full-size rafter (Fig. 8.5(c)) which, when cut, is used as a pattern for all the other rafters – ensuring uniformity across the whole roof.

Note: in practice, the cut at X is made *in-situ* after the rafters are all assembled. This gives a certain amount of leeway if any adjustments have to be made.

8.4 Roof assembly

Listed below is a sequential guide to the assembly of a short-span gable-ended single pitched traditional roof – to be used in conjunction with Fig. 8.6.

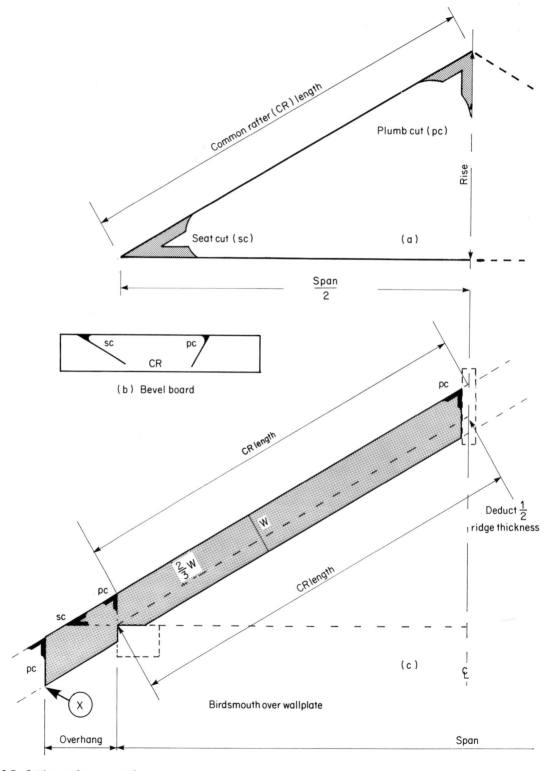

Fig. 8.5   Setting-out for pattern rafter

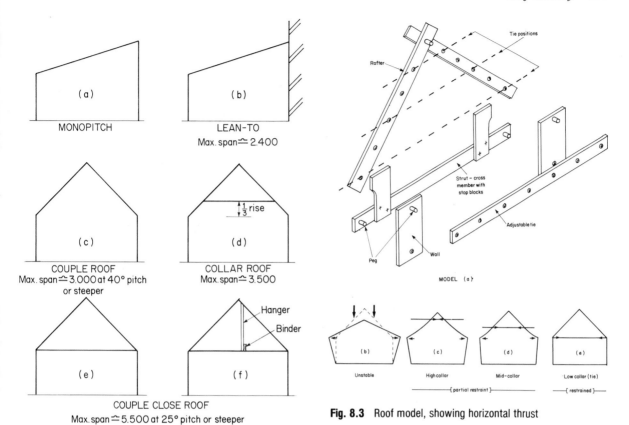

MODEL (a)

————{ partial restraint }———— ————{ restrained }————

Fig. 8.3 Roof model, showing horizontal thrust

MONOPITCH (a)

LEAN-TO (b)
Max. span ≃ 2.400

COUPLE ROOF (c)
Max. span ≃ 3.000 at 40° pitch
or steeper

COLLAR ROOF (d)
Max. span ≃ 3.500
⅓ rise

COUPLE CLOSE ROOF
(e) (f)
Max. span ≃ 5.500 at 25° pitch or steeper

Hanger
Binder

Fig. 8.4 Single roofs (pitched) for small spans

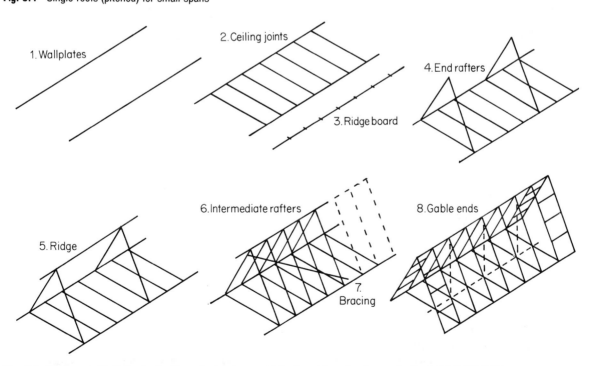

1. Wallplates
2. Ceiling joints
3. Ridge board
4. End rafters
5. Ridge
6. Intermediate rafters
7. Bracing
8. Gable ends

Fig. 8.6 Roof assembly Note: Roof will require bracing, and tying to gable ends, also anchoring down to the walls

1 *Wallplates*

  a) Assist the bricklayer to bed the wallplates
  and fix them (anchored down to the substructure) straight, level, and parallel.

  b) Couple close roof – mark the positions of
  ceiling joists, taking into account that their
  centres should relate to the size of ceiling
  material (if used).

  c) Couple and collar roof – mark the
  positions of rafters.

2 *Ceiling joists*   Position and nail the ceiling joists
to the wallplates – check that they are parallel.

3 *Ridge board*   Mark the positions of rafters from
the ceiling joists or wallplates.

4 *End rafters*   Position and nail two end pairs of
rafters to suit the length of the ridge
(scaffolding must be used to suit the situation).

5 *Ridge*   Position the ridge board (from
underneath). Nail the ridge to the rafters and
the rafters to the ridge at both ends.

6 *Intermediate rafters*   Fix enough rafters to enable
the roof framework to be plumbed and squared
(the number required usually depends on the
ridge-board length).

7 *Bracing*   Plumb the roof from one gable wall
and strap with a diagonal brace. Fix the
remaining rafters to suit the ridge-board
length.

8 *Gable ends*   Mark the position and fix the gable
ladder. Mark and fix binders and hangers if
required.

Note: Roofs require bracing, tying to gable
(end) walls, and anchoring down to their substructure – see *Carpentry and Joinery 2*.

If at any stage during the construction of a roof a
person is liable to fall more than 2 metres, a
working platform with suitable means of access
must be provided and used, in accordance with the
Construction Regulations 1966.

8.5 Eaves details

Flush eaves (Fig. 8.7(a))   Spar feet are cut about
25 mm longer than the outer face wall, to allow for
roof-space ventilation. Fascia boards are then
nailed to them to form a trim and provide a
bearing for gutter brackets.

Open eaves (Fig. 8.7(b))   Spar feet are allowed to
project well beyond the outer face wall. Fascia
boards are often omitted, the gutter being
supported by brackets fixed on the top or side of
the spar ends. Eaves boards mask the underside of
the roof covering.

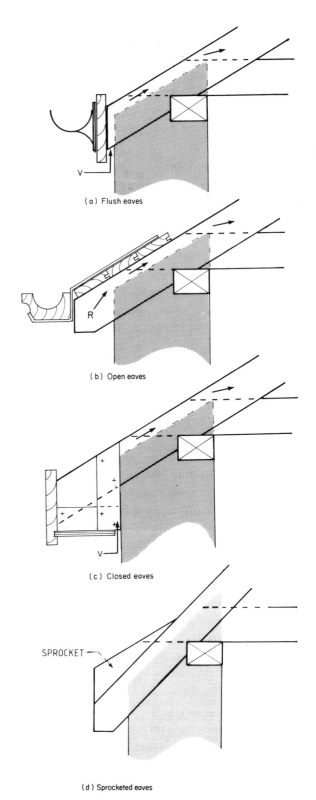

(a) Flush eaves

(b) Open eaves

(c) Closed eaves

SPROCKET

(d) Sprocketed eaves

Fig. 8.7   Eaves details V = ventilation to roof space, mesh-
covered to deter insects and birds etc.

Closed eaves (Fig. 8.7(c))   Spar feet overhang but are completely boxed-in (provision must be made for ventilation to roof space). Purpose-made brackets will be required to support the soffit at the wall edge. The front edge can be tongued into the fascia.

Sprocketed eaves (Fig. 8.7(d))   This is a method of reducing the roof pitch at the eaves of a steep roof, thus reducing the risk of water flowing over the gutter under storm conditions. The sprocket piece may be fixed on to or to the side of the rafter.

Note: Roof ventilation is dealt with in *Carpentry and Joinery 2*.

9

Turning pieces and centres up to 1 metre span

Where an opening is to be left in a wall as a passage, or to house a door or window frame, the

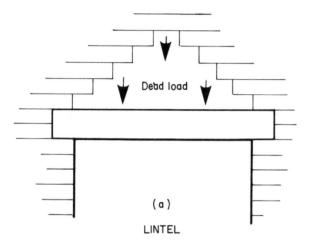

(a)

LINTEL

Load

Self-tightening

(b)

ARCH

Fig. 9.1 Support above an opening

load above it will need permanent support. A lintel (beam) can be used for this purpose, as shown in Fig. 9.1(a), or an arch can be formed. Arches allow the load above them to be transmitted around their shape as shown in Fig. 9.1(b).

The construction of an arch necessitates substantial temporary support until it has set. A centre not only provides this support but also provides an outline of the arch on which the bricklayer lays his bricks or the mason his blocks. Several arch outlines are shown in Fig. 9.2.

9.1 Simple geometrical arch shapes

Figures 9.3 to 9.7 illustrate methods of producing simple arch outlines. Arcs are scribed with the aid of a trammel bar and heads, as shown in Fig. 2.6, or by similar improvised means.

Semicircular arch (Fig. 9.3)

1 Bisect AB to produce point C.
2 Using radius R (CA), scribe a semicircle from point C.

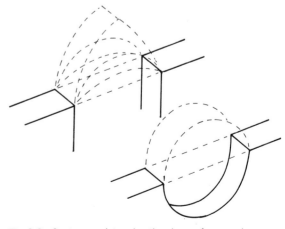

Fig. 9.2 Centres predetermine the shape of an opening

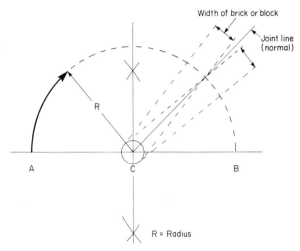

Fig. 9.3 Semicircular arch

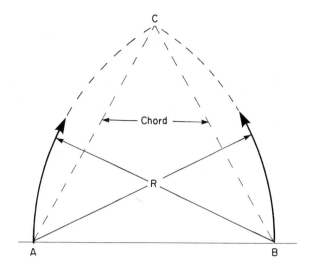

Fig. 9.5 Equilateral arch

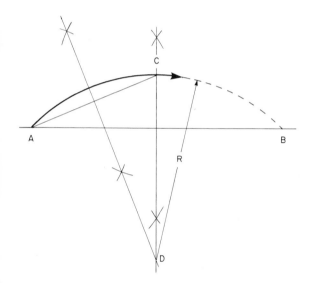

Fig. 9.4 Segmental arch

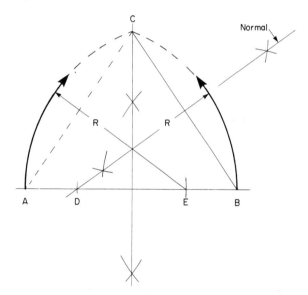

Fig. 9.6 Drop arch

Segmental arch (Fig. 9.4)

1 Bisect AB to produce a perpendicular line cutting AB.
2 Determine rise C.
3 Draw a line from A to C.
4 Bisect the chord AC to produce point D.
5 Using radius R (DA), scribe an arc from A through C to B from point D.

Equilateral arch (Fig. 9.5)

1 Using radius R (AB), scribe the arc BC from point A.
2 Similarly scribe arc AC from point B, point C being the intersection of both arcs.

Note: chords AC and BC will be the same length as span AB.

Drop arch (Fig. 9.6)

1 Bisect AB to produce a perpendicular line above AB.
2 Determine rise C – but ensure that chord BC is shorter than line AB.
3 Bisect chord BC to produce D on line AB.
4 Using radius R (DB), scribe arc BC from point D. Use the same radius to obtain E, by scribing an arc from point A.
5 Arc AC is then scribed from point E.

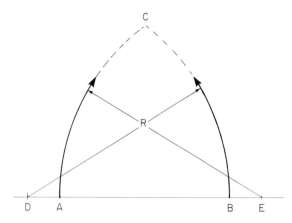

Fig. 9.7 Lancet arch

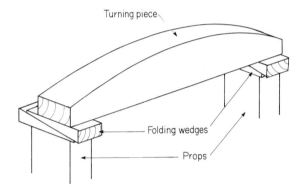

Fig. 9.8 Turning piece

Note: chords AC and BC will always be shorter than span AB.

Lancet arch (Fig. 9.7)

1 Using a distance longer than AB, scribe E on the base line from point A. Using the same distance, scribe D from point B.
2 Use radius *R* (DB or EA) from points A and B to locate C.

Note: chords AC and CB will always be longer than span AB.

The method used to produce the lancet arch can also be used for the drop arch, and vice versa.

9.2 Turning pieces

A turning piece is used where the rise and span of the arch is small. It consists of a single length of timber with its top edge shaped to suit the soffit (under-side) of the arch – it acts as a temporary

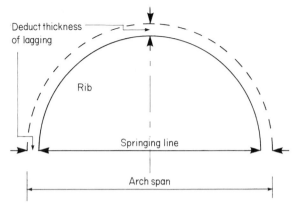

Fig. 9.9 Allowing for lagging

beam. Figure 9.8 shows a turning piece supported by props, with folding wedges as a means of adjustment, easing, and striking (see Section 9.4).

9.3 Centres

Centres are wooden structures built-up of the following members:

Ribs – form the profile of the arch and are made from sheet material (plywood) or solid sections joined with plywood or metal plates or are built-up of two thicknesses of timber with their joints lapping (see Fig. 9.13). Ribs provide support and fixing for lagging.

Lagging – battens or plywood, nailed on to the ribs to form a platform for the walling material. Lagging is termed either *closed* or *open*. Open lagging has spaces left between battens and is used with large stone or blockwork.

Ties – prevent built-up ribs from spreading and provide a fixing for bearers.

Struts – stabilise the framework by helping to redistribute some of the load placed on the ribs.

Bearers – tie the base of the centre and provide a sole, under which the centre is wedged and propped.

Note: all centres should be narrower than the wall thickness, otherwise they will hinder the bricklayer or mason when lining the wall through.

Construction

Start by drawing a full-size outline of half the centre. Remember to deduct the thickness of the lagging (except for centres for segmental arches) before setting out rib positions etc., as shown in Fig. 9.9.

Figure 9.10 shows a centre suitable for a segmental arch and how its width can be increased by adding an extra rib. Plywood has been used as close lagging.

Figure 9.11 shows a semicircular centre with plywood ribs and timber noggins, bearers, and lagging.

By using two or more ribs for each outline, larger spans become possible and more economical. The way ribs are joined gives them their identity, as mentioned above. A *solid* rib uses

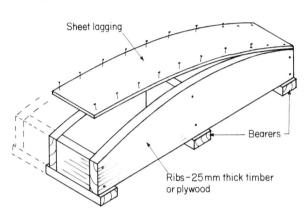

Fig. 9.10 Centre for segmental arch

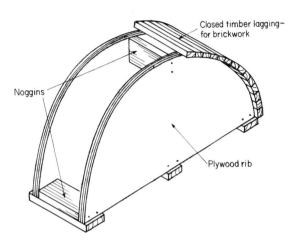

Fig. 9.11 Centre for semicircular arch

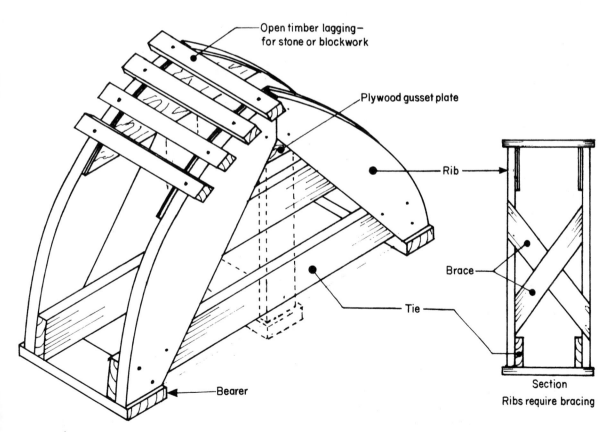

Fig. 9.12 Centre for equilateral arch

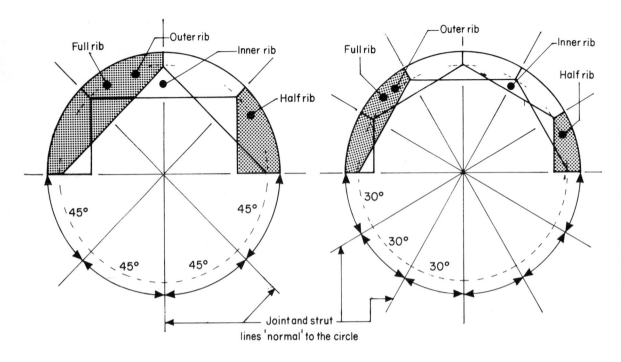

Note: the greater the number of ribs the more economical
use of materials, but more labour intensive

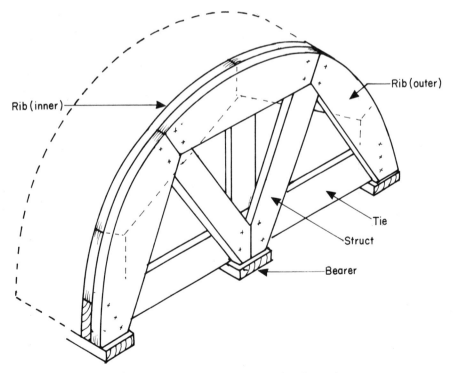

Note: a set of two laminated frames are required for each
centre–assembled similarly to Fig. 9.12

Fig. 9.13(a)   Built-up rib construction

plated butt joints as shown in the centre for an equilateral arch (Fig. 9.12).

Figure 9.13 shows a centre with a built-up rib and how a rib pattern enables ribs to be cut economically. Patterns can be cut from hardboard or thin plywood.

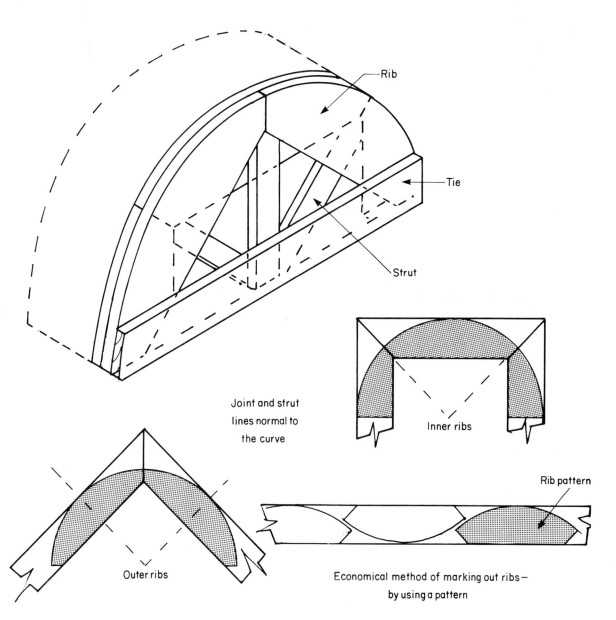

Rib

Tie

Strut

Joint and strut
lines normal to
the curve

Inner ribs

Outer ribs

Rib pattern

Economical method of marking out ribs—
by using a pattern

Fig. 9.13(b) Alternative built-up arrangement

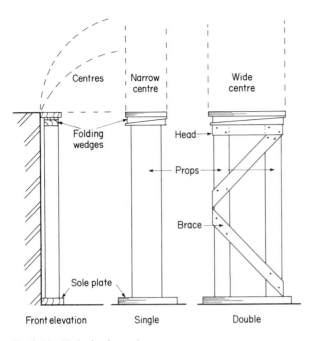

Centres

Narrow centre

Wide centre

Folding wedges

Head

Props

Brace

Sole plate

Front elevation

Single

Double

Fig. 9.14 Methods of propping

9.4 Easing and striking centres

Centres must be held in position without fear of their being displaced until the arch has thoroughly set, yet they must be capable of being gradually lowered. This lowering process, which takes place over a period of time, is known as *easing*. Folding wedges allow the centre to be lowered slowly – any sudden movement could damage the arch. Methods of propping narrow and wide centres are shown in Fig. 9.14. The eventual removal of the props and centre is known as *striking* and is made easy by using folding wedges.

10

Formwork

Formwork is best described as a temporary construction designed to contain wet concrete until it has set. There are two distinct types of formwork:

a) *in-situ work* – where concrete is *cast in-situ*, (*shuttering*) that is to say, in the position it is to occupy;

b) *pre-cast work* – used when concrete units are (*mould boxes*) *pre-cast* by forming them in a mould, either at a convenient position on site or under factory conditions (often by a firm which specialises in concrete products) to be used as and when required.

10.1 Formwork design

Because wet concrete is a heavy semi-fluid material, its formwork must be capable of restraining not only its mass but also its fluid pressure. The deeper the formwork, the greater the pressure it has to bear.

It is important to remember that the finished concrete surface will reflect the surface of its formwork. For example, sawn timber will produce a textured finish, just as planed timber, plywood (WBP), tempered hardboard, or sheet steel etc. will produce a smooth finish – provided their surfaces do not adhere to the concrete, which may be the case unless they are pre-treated with some form of parting agent or release agent. If a mould oil is used for this purpose, which also helps prevent air holes being left on the surface of the set concrete, its application must be strictly in accordance with the manufacturer's instructions.

For reasons of economy, formwork (particularly mould boxes) must be re-used many times, therefore initial design must include quick, simple, and in some cases mechanical methods of assembly and striking (dismantling) without undue damage, by such means as wedges, bolts, cramps, and nails

which can be easily redrawn without damaging the structure – preferably duplex-head types (Table 15.1).

The following items of formwork have been chosen for discussion because of their less complex nature. More advanced formwork will be featured in *Carpentry and Joinery 2*.

10.2 In-situ work (shuttering)

The simplest forms are those which provide only side and/or end support, such as for a concrete drive, path, or base for a light garage or shed etc.

Figure 10.1 shows a simple layout of formwork for a concrete patio or greenhouse base. Notice how a space has been left to one side to allow plants to be grown from the ground. Side boards are held by stakes firmly driven into the ground – packings between form board and stakes may be necessary where stakes are driven out of line. Stakes should be left long until the boards are fixed level. To avoid trapping encased boards, the corners have been cut back.

10.3 Pre-cast work (mould boxes)

The shape of even a simple moulded object such as a jelly or a sand castle will hopefully reflect that of the mould from which it was produced. Similarly, concrete items such as those shown in Fig. 10.2 should also accurately reproduce the shape of their respective moulds.

Because of the weight of their contents, or the need to cast several items together (Fig. 10.3), mould boxes used for concrete products are usually made to dismantle easily.

Figure 10.4 shows in orthographic projection how a double mould box could be constructed for a stump post, suitable for anchoring fence posts etc. to the ground. The holes in the post are formed by casting plastics or cardboard tubes into the

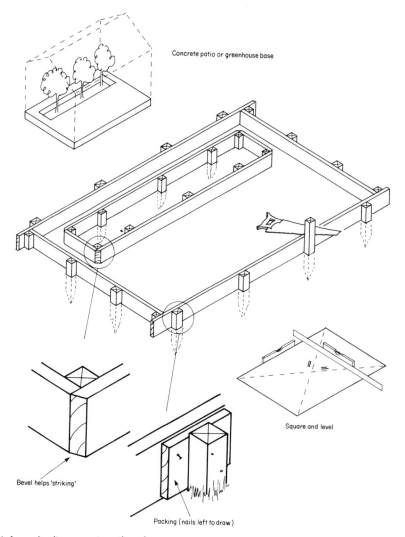

Concrete patio or greenhouse base

Square and level

Bevel helps 'striking'

Packing (nails left to draw)

Fig. 10.1   Formwork for an in-situ concrete patio or base

Concrete items

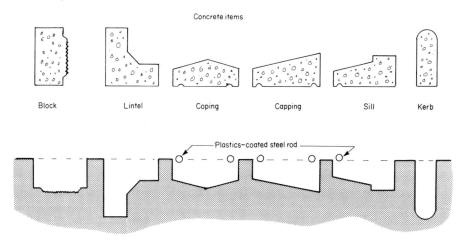

| Block | Lintel | Coping | Capping | Sill | Kerb |

Plastics–coated steel rod

Fig. 10.2   Common pre-cast
concrete products

CASTING POSITIONS

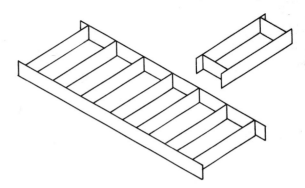

Fig. 10.3 Single and multiple casting units

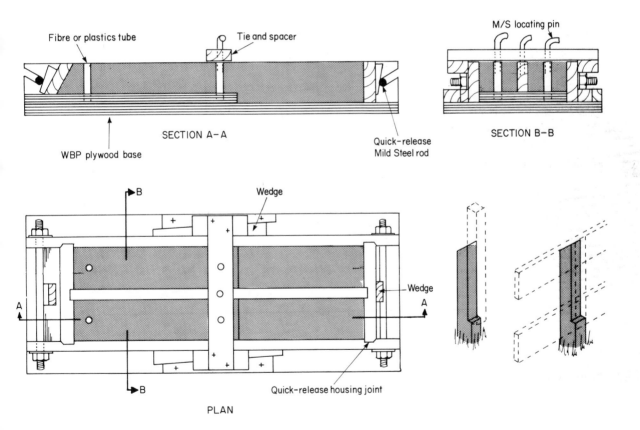

Fig. 10.4 Double mould for
pre-cast concrete posts

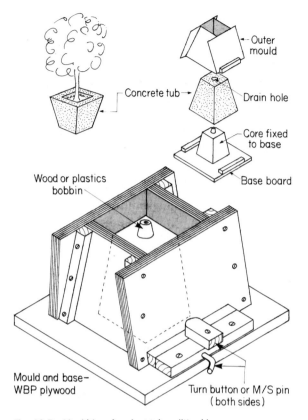

Outer mould

Concrete tub

Drain hole

Core fixed to base

Base board

Wood or plastics bobbin

Mould and base– WBP plywood

Turn button or M/S pin (both sides)

Fig. 10.5   Mould box for plant tub or litter bin

concrete. Notice how the bolts which hold the box sides in position can be slid out without removing the nuts, and how the end-stop housings are bevelled for easy seating and removal.

The mould for a plant tub or litter bin shown in Fig. 10.5 is kept intact, because the shape of its outer shell allows it to be lifted off, leaving the casting on the core mould until the concrete is cured (completely hardened).

11

Ledged-and-braced battened (matchboarded) doors

This type of door (door leaf) is commonly used as an exterior door to outhouses, sheds, garages (for pedestrian access), and screens. Provided it is made and fixed correctly, it will withstand a lot of harsh treatment and remain serviceable for many years.

11.1 Door construction

As can be seen from Fig. 11.1, the face of the door is made up of V or beaded tongue-and-groove matchboard, known as *battens*. These are nailed or stapled to three horizontal members called *ledges*, these ledges being held square to the battens by

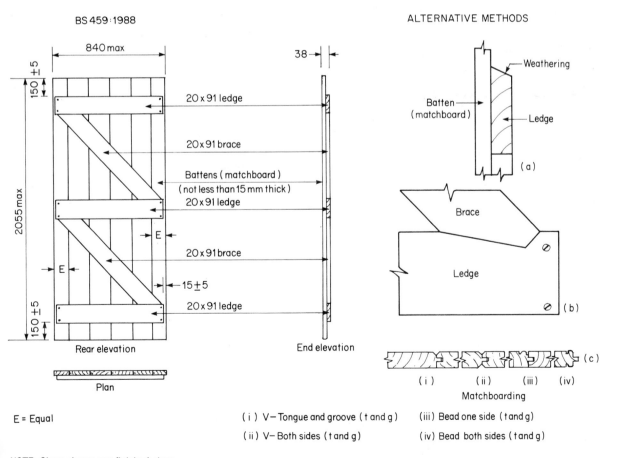

E = Equal

(i) V– Tongue and groove (t and g)

(ii) V– Both sides (t and g)

(iii) Bead one side (t and g)

(iv) Bead both sides (t and g)

NOTE: Sizes shown are finished sizes

Fig. 11.1 Ledged-and-braced battened (matchboarded) door construction

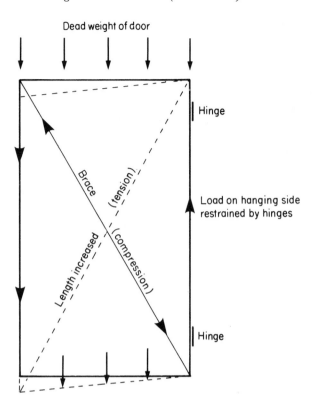

Dead weight of door

Hinge

Load on hanging side
restrained by hinges

Hinge

– – – Indicates possible movement
when the brace is dispensed with

(a)

Fig. 11.2(a)   Bracing principle

(b)

Fig. 11.2(b)   How a ledged-and-battened (matchboarded) door
should not be fixed
Note: widening gap at the top of the door (due to omission or
wrongly positioned brace). Also the 'thumb latch' closing
mechanism should not have been used in this situation - the
'beam' is on the wrong door face (see Fig. 11.9)

similarly fixing two diagonal pieces of timber called
braces.

A less expensive version of this door is the
ledged-and-battened door. As its name implies, it
is built without a brace – its stability therefore
relies entirely upon its nailed contruction, which
under normal circumstances would prove
inadequate. It can, however, be used quite
satisfactorily in small openings or as a temporary
door.

Door braces

The importance of a brace is illustrated in
Fig. 11.2(a), where it will be seen that its
effectiveness is controlled by its direction: if the
door is to retain its shape, the brace *must* always
point away and in an upwards direction from the
hanging side (hinged side), otherwise the door
could sag at the closing side (Fig. 11.2(b)) – hence

the term *sag bar* sometimes used when referring to a
brace.

The making of a simple card or wood model
(Fig. 11.3) should help to clarify bracing
principles:

1   Cut four pieces of stiff card, thin plywood, or
hardboard etc. 500 mm long by 50 mm wide to
form the sides, top, and bottom.
2   Join them together at the corners with a single
pin, nail, or screw. This will allow each corner
to pivot and produce a scissor movement.
3   Lay the frame flat. Move the corners until they
are square (at 90° to each other).

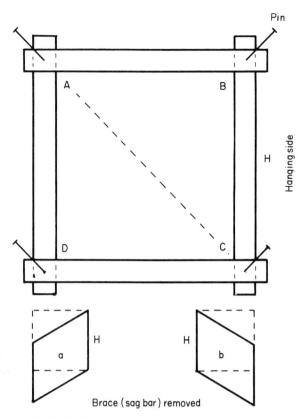

Fig. 11.3 Model to stimulate frame movement

4 Cut a piece of rigid material to fit between corners A and C – this will act as the brace.

5 While the brace is held in this position, take hold of side H. Lift the whole frame up and turn it until vertical. The brace should now be self-supporting, and therefore the frame will remain square. However, if the bracing piece had been omitted, or positioned into the opposite corners B and D, the frame would have collapsed. In fact, corners B and D would have become wider apart (Figs 11.3(a) and (b)), so allowing the brace to fall out.

BS 459:1988

Figure 11.1 shows some of the requirements specified by BS 459:1988 for the construction of a ledged-and-braced battened door, but it is worth noting that traditionally such doors have been and are still in some cases made from heavier sectioned timber and may also include other features such as those shown in Fig. 11.1 – for example, weathering to ledges (bevelling on their top edge) where the door is subjected to weather from both

sides (Fig. 11.1 (a)), or extra brace restraint by housing braces into the ledges (Fig. 11.1(b)).

11.2 Assembly

A typical order of assembly for a ledged-and-braced battened door is illustrated in Fig. 11.4.

Stage 1

a) Cut the battens and ledges to length.
b) Check that the ledges are not twisted (plane them out of twist if necessary).
c) Paint or preserve all the tongues and grooves and the ledge faces that will come into contact with the battens.
d) Rest the battens face down across the bearers, then lightly cramp the battens together.
e) Double screw each ledge to the cramped *edge* battens (Fig. 11.5).

Stage 2

a) Cut the braces to fit between the ledges.
b) Edge nail the braces to the ledges (ensure that the door is kept square during this operation – if necessary, use end stops).

Stage 3

a) Turn the door over on to bearers which have been placed lengthwise.
b) Double nail each batten to the ledges and braces, taking care to avoid the bearers, as after the nails have been punched below the surface they will protrude 10 mm through the door.

Stage 4

a) Turn the door back on to its face, where the protruding point of each nail will be visible.
b) Clench each nail, by bending it over in the direction of the grain, then punch the clench below the surface of the wood.

11.3 Framing

If the door is to be used to close an opening in a brick or block structure, a firmly fixed timber frame will be needed to support the door when hung (swung on hinges) and provide a means of making it secure.

The door frame will require a rebate to act as a door check, otherwise the door would swing through the opening, straining the hinge and/or

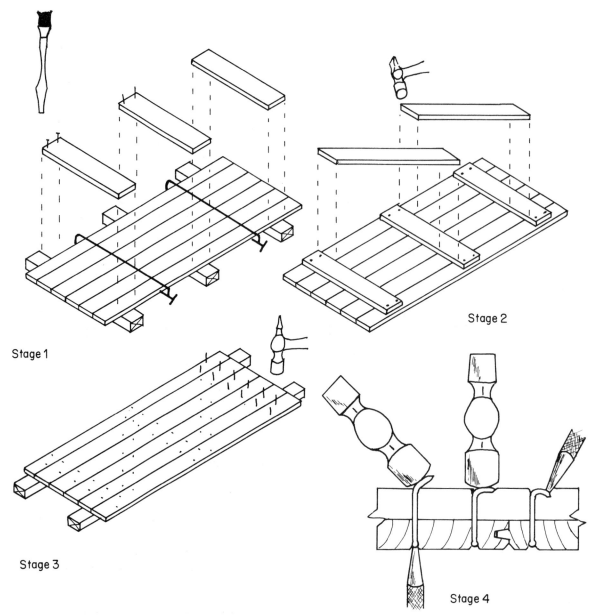

Stage 1

Stage 2

Stage 3

Stage 4

Fig. 11.4 Stages of door assembly

splitting the jamb. Methods of forming a rebate can be seen in Fig. 11.6(a). The width of the rebate will depend on whether the door is to swing outward or inward, i.e. whether the hinges are screwed directly on to the ledges or via the matchboard (Fig. 11.6(b)).

Figure 11.7 deals with the making, assembly, and fixing of a suitable door frame. The head and jambs are joined together by using a mortise-and-tenon joint (Figs 11.7(a) and (b)) which should be

coated with paint or a suitable resin adhesive, assembled, cramped, wedged, and dowelled. If cramping the joint is not practicable, the joint could be draw-bored (Fig. 11.7(c)) – when a hardwood dowel is driven through the off-centre holes, the shoulders of the joint will be pulled up tight. Because the frame has only three sides, a temporary tie (distance piece) fixed across the bottom of the jambs and a brace at each corner will be necessary if it is to retain its shape.

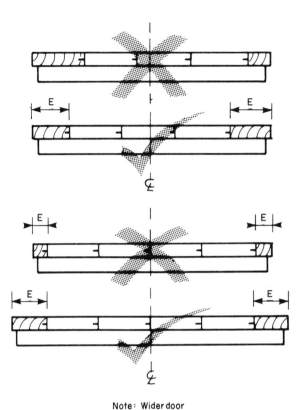

Fig. 11.5 Positioning edge battens. NB: Edge battens must be equal (E) and of sufficient width to allow for fixing - achieved by placing either a batten or joint to the centre of each ledge

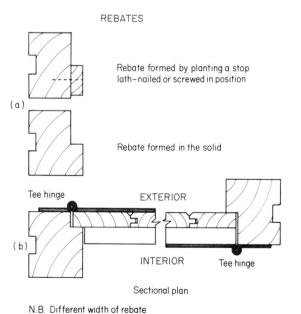

Fig. 11.6 Forming rebates and positioning the door

If the frame is to be built into the structure, temporary propping will be needed (Fig. 11.7(d)). When the frame has been accurately positioned – both level and plumb – the permanent securing process can begin. Firstly, provision is made at the foot of each jamb for good anchorage to the step, either by using a metal dowel (Fig. 11.7(e)), shoe (Fig. 11.7(f)) or a concrete stool (Fig. 11.7(g)). As the walls are built up on either side, wall clamps are fixed at approximately 500 mm intervals with screws to the back of the jambs (Fig. 11.7(h)), then walled in. On reaching the head (Fig. 11.7(a)), it will be seen that the *horns* have been cut back on the splay. This allows the face walling to lap and totally enclose them, to make ready for the lintel above.

If, however, the frame is to be fixed into an existing opening, other fixing devices will have to be employed, for example wood or plastics plugs, wedges, etc. (see Chapter 15). Specially formed openings should have built-in fixings – timber blocks (pallets) sandwiched between the mortar and courses of brick or blockwork, etc. Either way, it follows that the horns will not be needed as a fixing aid and they will therefore have to be sawn off. For greater stability, a haunched mortise-and-tenon joint is then used (Fig. 11.7(b)).

11.4 Fitting and hanging

1 Check that the bottom of the door is parallel with the step or threshold. Fit it if necessary.
2 Fit the door's hanging edge into its rebate.
3 a) With an assistant, position the door's hanging edge just inside its rebate. Using half the width of a pencil, scribe the closing edge. This will provide door clearance (Fig. 11.8). Remove waste wood and bevel back to produce a 3 mm lead in – known as a *leading edge*.
   b) Alternatively, measure the distance between rebates (allowing for clearance), transfer it to the door's face, and remove waste wood as in 3(a).
4 Place the door into the rebates, remembering to allow for floor clearance. Mark its height from inside the top rebate and remove waste wood.
5 Lay the door flat over two saw stools and screw tee hinges (Fig. 11.9) to the battens or ledges, using only two screws per hinge.
6 Position the door into its frame. Use thin wedges as packings and adjust to give clearance all round (2–3 mm at the sides and

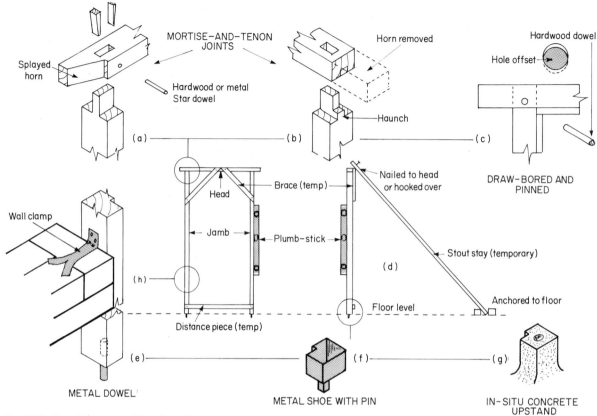

Fig. 11.7   Door frame assembly and erection

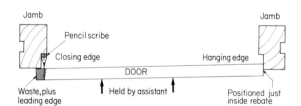

Fig. 11.8   Fitting the door to the closing edge

top, 4–6 mm to the floor). Fix the top hinge to the jamb, using only one screw, then the bottom likewise.

7   Remove the packings. Check the clearance, noting any adjustments needed, and remedy if necessary.

8   Repack the door. Unscrew the door from its frame (bottom hinge first) then remove both hinges from the door.

9   Paint the backs of the hinges and refix them to the door, using *all* the screws this time.

10   Rehang the door and fix all the remaining screws.

Note: when using pre-painted hinges, stages 5–10 will not wholly apply.

Hardware (ironmongery)

All that remains now is the choice of hardware to provide a way of keeping the door closed, e.g. a latch, lock, or bolt. The door's use, location, and accessibility will be the deciding factors here. If the door is simply to be held closed and to be opened from both sides, then a thumb latch would be an ideal choice. If, on the other hand, the door has to be fixed shut at some time, for security reasons, then the addition of a lock and/or bolt should be considered.

The thumb latch, sometimes known as a Norfolk or Suffolk latch, provides a simple yet trouble-free means of holding the door closed and should be fixed in the following manner:

1   Make a slot in the door, into which the sneck is pushed.

2 Screw the face-plate (to which the sneck is hinged) to one face of the door.
3 Fix the beam and keeper to the opposite side.
4 The stop can now be screwed to the frame.

Note: black japanned round-headed screws should be used throughout.

Probably the simplest way of making the door secure is by the use of a tower bolt, screwed to the door's ledge. A press lock or rim deadlock will require a keyhole to be made in the door (Fig. 11.9(a)).

11.5 Items and processes

An overall sequence of operations can be listed in chart form as an easy means of reference. Table 11.1 shows how the door's construction can be tabulated.

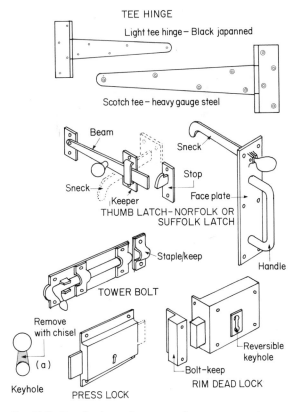

Fig. 11.9 Door hardware (ironmongery)

Table 11.1 Sequence of operations for door construction

Process	Item	Member	Operation	Machines	Powered hand tools	Hand tools	Ironmongery	Remarks
Preparing timber	Door	Ledges, battens, braces	Cut to length, ripping, deeping.	Cross-cut circular-saw bench	-	-	-	Cut to nominal size - ex.
			Planing	Hand-feed planer/surfacer, panel planer/ thicknesser	-	-	-	Finished size - width and thickness
		Battens	Form V, tongue and groove	Spindle moulder	Router	Rebate plane plough plane	-	Worked by machine or hand
Assembly	Door	Battens, ledges, braces	Fix battens to ledges		Staple gun	Hammer	Nails/staples	Paint or preserve T & Gs - leave to dry before assembly.
			Cut bevel and fix	-		Panel saw, hammer		
			Punch and clench nails			Hammer, nail punch		Boards held face to face during treatment.
Finish	Door		* Sanding	Belt sander	Belt/orbital sander	Cork sanding block, glass or garnet paper	-	Finish depends on grade of abrasive * usually before assembly
	Door		Knotting Painting	-	-	Paint brush	-	Shellac knotting Priming paint

12

Single-light casement windows

The main purpose of a window is to allow natural light to enter a building yet still exclude wind, rain, and snow. It also serves as a means of providing ventilation if the glazed area is made to open. An openable *light* (a single glazed unit of a window) is called a *sash* or *casement* – hence the term *casement window*.

Figure 12.1 shows details of two types; the *traditional*, which houses its sash fully within its frame, and the more modern *stormproof* type which, because of its double-rebate system (i.e. both the frame and the sash are rebated) gives better

weather protection and increased natural light. Because the rebate in the frame of the stormproof type is only half the sash width, direct glazing (a deadlight) can be incorporated into the frame.

12.1 Construction

Construction methods will depend mainly on the type of window and the sectional profile of its members. Traditional sections (and stormproof sections to BS 644) are shown in detail in Fig. 12.1. Figure 12.2 shows how these sections

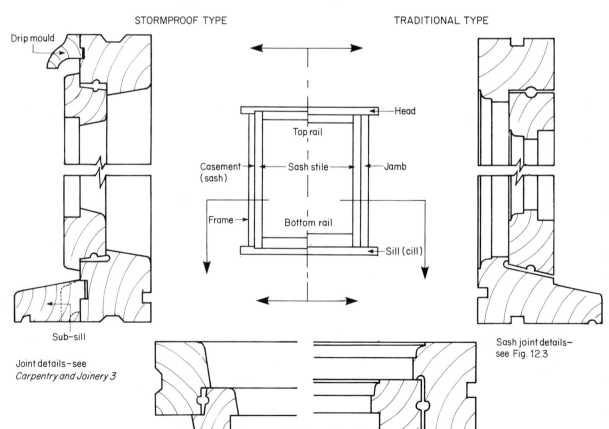

Fig. 12.1 Sectional details of single-light casement windows

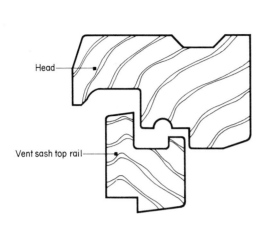

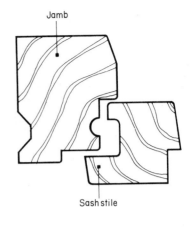

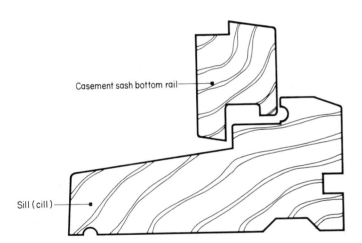

Fig. 12.2 Examples of how window sections may be modified

can be modified. The important feature of all these sections is how the grooves are formed to provide protection against the entry of moisture by capillary action (see Chapter 13).

Window frames

Frames should be held together at the corners with either combed joints (multiple corner bridles) with not less than two tongues, or mortise-and-tenon joints. The tongues or tenons should be not less than 12 mm in thickness. Where horns are required for building-in purposes, they should not be less than 40 mm in length. Joints should be glued, with a suitable synthetic adhesive, and be

held by wedges and/or pegged with hardwood or metal star dowels (see Fig. 11.7(a)).

Casements (sashes)

The corners should be joined by using combed or mortise-and-tenon joints, glued (as above), and pegged and/or wedged (depending on the type – bridle joints cannot be wedged). Traditional casement joints are shown in Fig. 12.3. Storm proof casement details are featured in Book 3.

Sash hanging

Sashes can be hung in one of the three ways shown in Fig. 12.4, where the apex of the V indicates the hinging side and that the knuckle (Fig. 12.5) of the

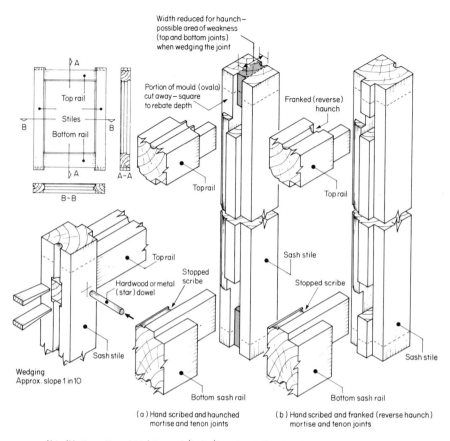

Width reduced for haunch – possible area of weakness (top and bottom joints) when wedging the joint

Portion of mould (ovalo) cut away – square to rebate depth

Franked (reverse) haunch

Top rail

Top rail

Stiles

Bottom rail

Top rail

Top rail

Sash stile

Top rail

Stopped scribe

Hardwood or metal (star) dowel

Stopped scribe

Sash stile

Wedging
Approx. slope 1 in 10

Sash stile

Bottom sash rail

Bottom sash rail

(a) Hand scribed and haunched mortise and tenon joints

(b) Hand scribed and franked (reverse haunch) mortise and tenon joints

Note (Mortise and tenon joints) the use of a 'franked' haunch on <u>shallow</u> rebate/moulded small sectioned stock reduces the risk of splitting the stile during wedging and can produce a stronger joint

Fig. 12.3   Joints using traditional sash stile and sash rail sections

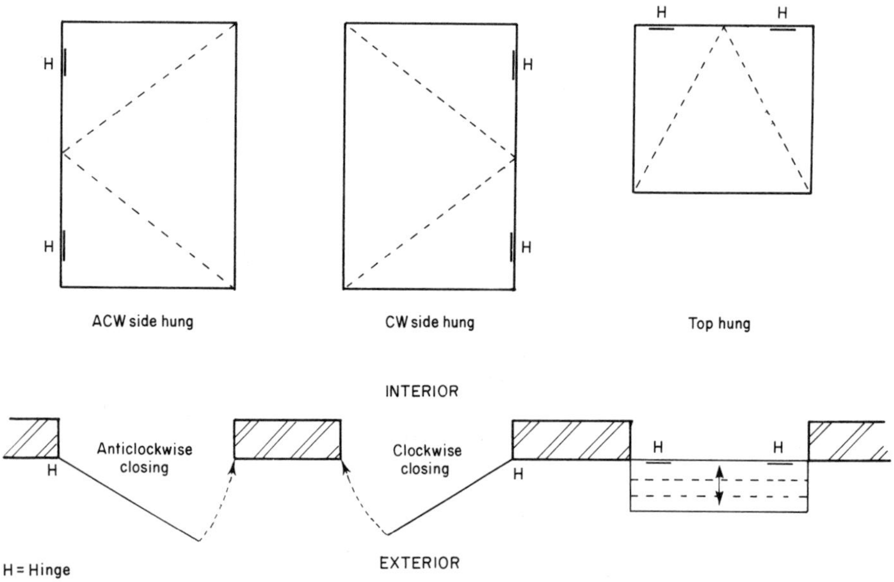

H

H   H

H

H

H

ACW side hung

CW side hung

Top hung

INTERIOR

Anticlockwise closing

Clockwise closing

H

H

H

H

EXTERIOR

H = Hinge

Fig. 12.4   Opening casements

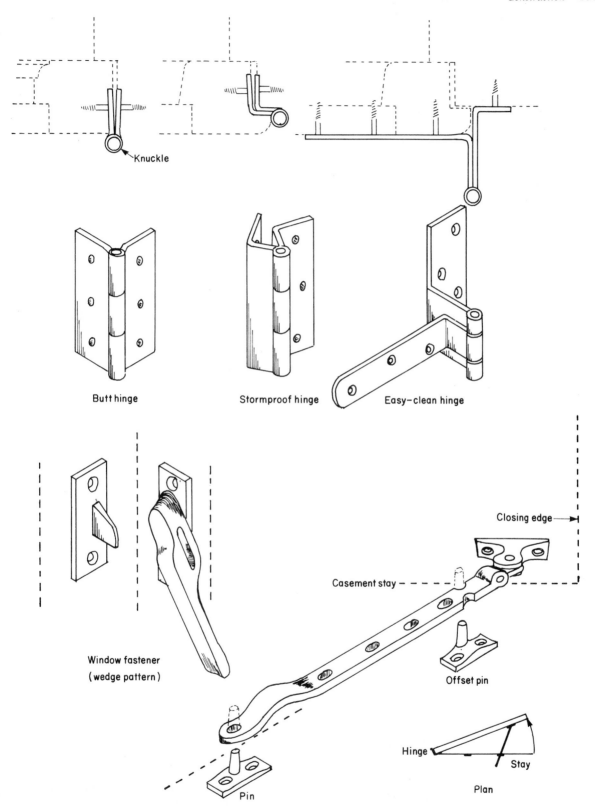

Fig. 12.5 Window hardware (ironmongery)

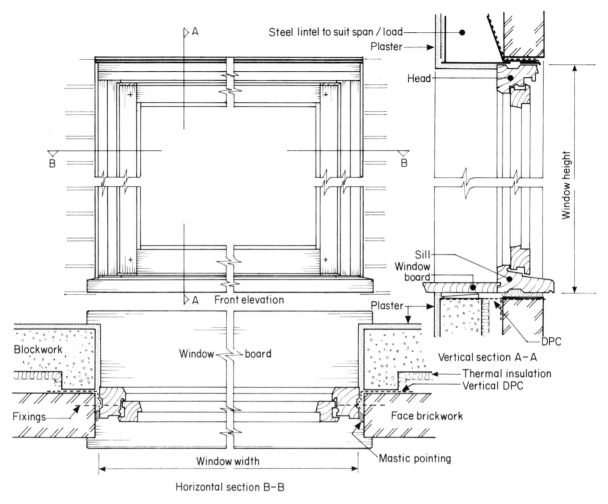

Fig. 12.6   Window installation details

hinge is to that face. Three different types of hinge are shown in Fig. 12.5, together with their respective fixing details.

The screws used to fix the hinges will be exposed to varying amounts of strain, particularly those used on the top hinges of side-hung sashes, and those on sashes which are top hung which will constantly be subjected to *withdrawal*.

Because openable sashes must provide varying amounts of ventilation, a multi-position casement stay similar to the one shown in Fig. 12.5 is used. This also gives a means of securing the sash when closed. Side-hung sashes also use a window fastener (Fig. 12.5) fixed midway to the closing-side stile and jamb.

Glass is not usually fixed until the window is built into the structure, but its weight must be

taken into account with regard to hinge and screw sizes. The glass also plays an important part in holding the sash square.

12.2 Fixing to the structure

Window frames can either be built into the fabric of the structure as building progresses, in a similar manner to the door frame shown in Fig. 11.7, or be fixed with screws and plugs into a pre-formed opening after the main structure is built. The former is the most common method used in house building.

Figure 12.6 shows a vertical section of how a typical window unit is incorporated into the main outer fabric of a house.

13

Moisture movement

One of the main requirements in building a habitable dwelling is that moisture be excluded from its interior. Failure to do so could result in a high moisture content of timber and its eventual breakdown by moisture-seeking fungi, not to mention the many other structural and environmental effects associated with dampness.

Most building materials are porous (contain voids or pores), thus allowing moisture to travel into or through them. Moisture may enter these materials from any direction, even the under-side and travel upwards as if defying the laws of gravity.

Upward movement of moisture is responsible for rising damp and is caused by a force known as a *capillary* force which in the presence of surface tension produces an action called *capillarity*.

13.1 Surface tension

The microscopic molecules of which water is composed are of a cohesive (uniting or sticking together) nature which, as a result of their pulling together, produces the apparent effect of a thin flexible film on the surface of water. This effect is known as *surface tension*.

The reluctance of this *skin* to be broken can be seen by filling a container with water just below its brim, then slowly adding more water until a *meniscus* (curved surface of liquid) has formed (Fig. 13.1(a)). Its elasticity can be demonstrated further by carefully floating a small sewing needle or a thin flat razor blade (double-edged safety-razor type) on its surface. Figure 13.1(b) shows how a depression is made in the 'skin' by the weight of a needle. If, however, surface tension is 'broken' by piercing the 'skin', the needle will sink.

13.2 Capillarity

A clear clean glass container partly filled with water will reveal a concave meniscus where the water has been drawn up the sides of the glass (Fig. 13.2(a)). This indicates that a state of adhesion (molecular attachment of dissimilar materials) exists between the glass and water molecules and at this point is stronger than the cohesive forces within the water.

Figure 13.2(b) shows what happens when a series of clean glass tubes is stood in a container of water. The water rises highest in the tube with the smallest bore (hole size), indicating that the height reached by capillarity is related to the surface area of water in the tube. It can therefore be said that, if the surface area of water is reduced or restricted, the downward pull due to gravity will have less effect, whereas upward movement will be encouraged by surface tension.

A similar experiment can be carried out by using two pieces of glass as shown in Fig. 13.2(c). This method produces a distinct meniscus which can be varied by making the gap at the open end wider or narrower.

It should now be apparent that, for a capillary force to function, all that is needed is water with its

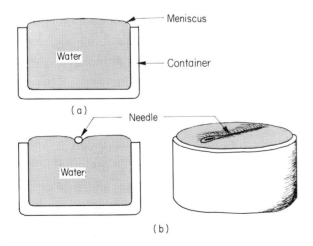

Fig 13.1 The effect of surface tension

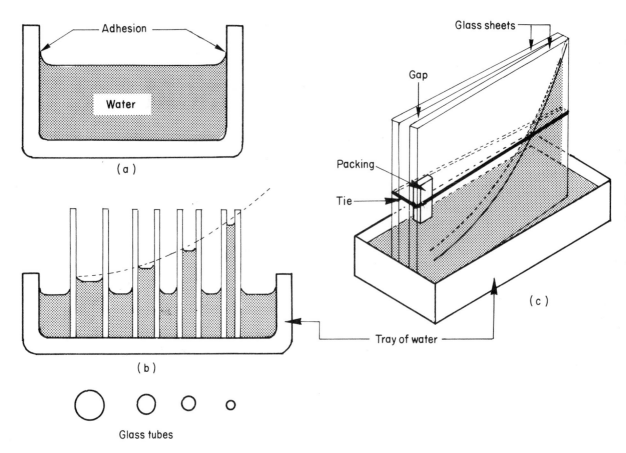

Fig 13.2   The effect of capillarity

cohesive properties and a porous material, or a situation providing close 'wettable' surfaces which encourage adhesion.

13.3 Preventative measures

A knowledge of how water acts in certain situations and reacts towards different materials enables a building and its components to be designed with built-in water and moisture checks. For example, the common method of preventing moisture from rising from the ground into the structure is by using a horizontal damp-proof course (DPC) like those shown in Fig. 7.1, where penetrating damp has also been avoided by leaving a 50 mm cavity between outer and inner wall skins. A vertical DPC would be used where both skins meet around door and window openings.

There are, however, many other places –

particularly the narrow gaps left around doors and sashes – which would provide ideal conditions for capillary action if preventative steps were not taken. The first line of defence should be to redirect as much surface water away from these areas as possible, by recessing them back from the face wall and encouraging water to drip clear of the structure.

Figure 13.3 shows how a 'throat' (groove) cut into the underside of overhangs (i.e. thresholds, window sills, drip moulds, etc.) interrupts the flow of water by forcing it to collect in such a way that its increase in weight results in a drip being formed. Further examples of throatings are shown in Figs 12.1, 12.2, and 12.6.

It is inevitable that water will find its way around those narrow gaps, so anti-capillary grooves, or similar, are used. Figure 13.4 shows how these grooves work, and Figs 12.1 and 12.2 illustrate how they are included in window design.

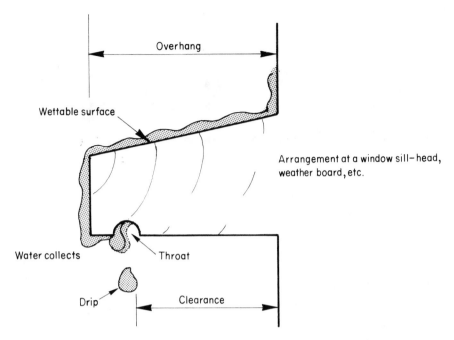

Fig 13.3   Encouraging water to drip

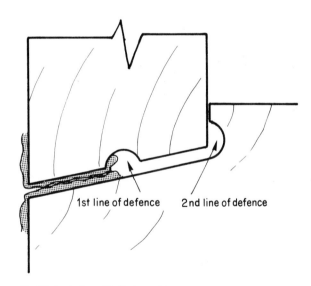

Fig 13.4   Anti-capillarity grooves

14

Shelving

Shelves are expected to support the differing weight-to-volume ratios of a vast variety of items and materials. It is therefore essential that, before shelves are constructed or assembled, consideration is given to both their function and location, because these are the essential factors that must be integrated within the overall design.

Figure 14.1(a) shows the dangers of

1 using shelving of inadequate thickness or strength to withstand the necessary imposed loads, or
2 expecting shelving to span an unrealistic distance.

Figures 14.1(b) and (c) can be regarded as suitable remedial alternatives, i.e. providing the necessary intermediate support and/or using stronger material.

Shelves are made up of either *solid* or slatted material, and examples are shown in Fig. 14.2 (see also Section 6.2).

Over the years, various methods of supporting shelves have evolved which the joiner can adapt to suit his specific requirements. A few examples of these methods will be found in Sections 14.1 and 14.2. You should also consult Chapter 15 with regard to fixing devices.

14.1 Traditional shelving

As can be seen from Fig. 14.3, shelves can be supported by

a) bearers fixed to a wall,
b) brackets – metal, or purpose-made from timber and/or plywood,
c) timber-framed uprights (standards),
d) solid uprights (enclosed units), where provision for shelf adjustment can be made.

The methods in Figs 14.3(a) and (b) rely on walls for their support, whereas those in Figs 14.3(c) and (d) may be free-standing units.

14.2 Proprietary systems and shelving aids

Commercially produced forms of shelf support may follow a similar pattern to those in Figs 14.3(b) and (d) but provide for greater flexibility in both their construction and the means of shelf adjustment.

Figure 14.4(a) shows how metal brackets of various styles and sizes can be positioned to any height by hooking them on to an upright metal channel which has been secured to a wall. Figure 14.4(b) shows how a series of holes bored in a solid upright enables metal or plastics shelf-supporting studs and their sockets to be positioned to suit shelf requirements. Alternatively, a metal strip can be housed or surface-fixed with adjustable pegs as shown in Fig. 14.4(c).

Conclusion Stop for a moment to consider how shelves are utilised around your home and place of work, then ask yourself: Are they suitable for the

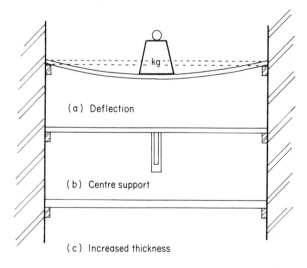

(a) Deflection

(b) Centre support

(c) Increased thickness

Assume shelf end supports are adequate

Fig 14.1 Shelf support

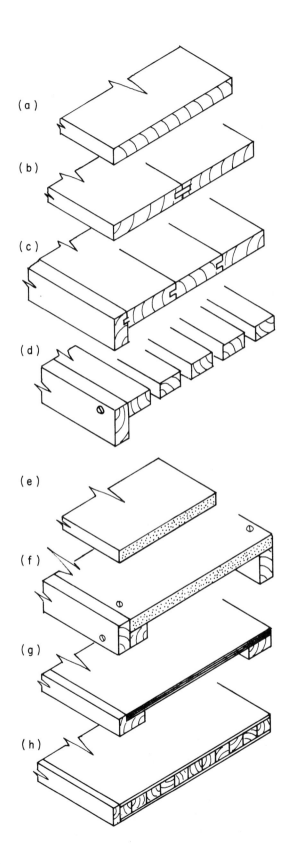

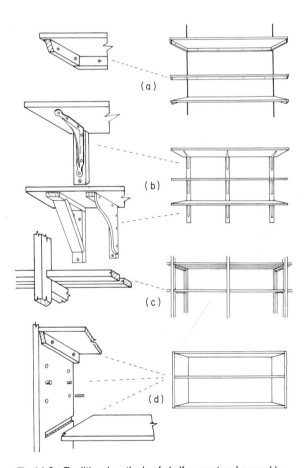

Fig 14.2 Shelf boards and their construction: (a) Single board; (b) double board with loose tongue; (c) tongued-and-grooved boards with front stiffener; (d) slatted with front stiffener; (e) particle board — self finish or with veneered wood or plastics; (f) particle board with front and back stiffeners; (g) plywood with front stiffener and back support; (h) blockboard or laminboard with slipped front edge

Fig 14.3 Traditional methods of shelf support and assembly

job they are doing? Are they strong enough? Should they be wider or higher? Are the method of construction and the shelf material in keeping with their surroundings? etc. From your answers, you should be able to draw your own conclusion as to the importance of this chapter.

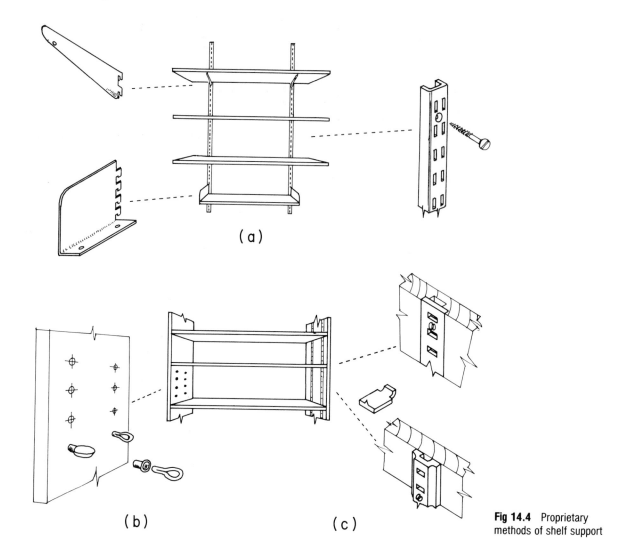

Fig 14.4 Proprietary methods of shelf support

15

Fixing devices

The decision as to how a piece of timber is fixed and which device to use is usually left to the joiner – unless the designer states otherwise. In either case, the following factors should be considered:

1 location,
2 strength requirements,
3 resistance to corrosion,
4 appearance,
5 availability,
6 cost.

15.1 Nails

Nails offer the quickest, simplest, and least

Table 15.1 Nails, pins, and staples

Nail type	Material	Finish or treatment	Shape or style	Application
Wire nails				
Round plain-head	S	SC		Carpentry; carcase construction; wood to wood
Clout (various sized heads)	S, C	SC, G		Thin sheet materials; plasterboard; slates; tiles; roofing felt
Round lost-head	S	SC		Joinery; flooring; second fixing. (Small head can easily be concealed.)
Oval brad-head	S	SC		Joinery; general-purpose. Less inclined to split grain.
Oval lost-head	S	SC		As for Oval brad-head.
Improved nails				
Twisted shank	S	SC, G		Roof covering; corrugated and flat materials, metal plates, etc.;
Annular-ring shank		SH		flooring; sheet materials. Resist popping (lifting). Good holding power, resisting withdrawal.
Duplex head	S			Where nails are to be re-drawn - formwork etc.
Cut nails				
Cut clasp	S	SC		Fixing to masonry (light weight (aerated) concrete building blocks etc.) and carpentry. Good holding properties.
Flooring brad (cut nail)	S	SC		Floor boards to joists (good holding-down qualities)
Panel pins				
Flat head	S	SC, Z		Beads and small-sectioned timber
Deep drive		G		Sheet material; plywood; hardboard
Masonry nails	S/hardened and tempered	Z		Direct driving into brickwork, masonry, concrete. (Caution: not to be driven with hardened-headed hammers. Goggles should always be used.)
Staples (mechanically driven)	S	Z	Temporarily bonded	Plywood Fibreboards Plaster-boards Insulation board to wood battens
Corrugated fasteners ('dogs')	S	SC	Joint	Rough framing or edge-to-edge joints
Star dowel	A	SC		An alternative to hardwood dowel for pinning mortise and tenons or bridles

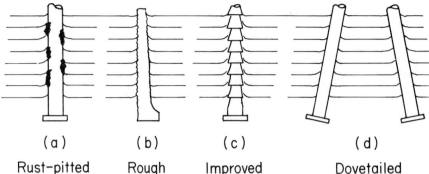

(a) Rust-pitted   (b) Rough   (c) Improved   (d) Dovetailed

Fig 15.1 Nails - resistance to withdrawal

expensive method of forming or securing a joint and, provided the material being fixed is suitable and the nails are the correct size (length approximately $2\frac{1}{2}$ times the thickness of the timber being fixed), the correct shape, and correctly positioned to avoid splitting, a satisfactory joint can be made.

Table 15.1 illustrates different types of nails, pins, and staples and describes their common use.

When dismantling a joint or fixture which has been nailed, pay particular attention to the effort required to withdraw the nail, and whether the head pulls through the material. The ease or difficulty of withdrawal emphasises the importance of choosing the correct nail for the job. Apart from the nail size or type of head, resistance to withdrawal could be due to the following;

a) type of wood,
b) rust and pitting (Fig. 15.1(a)),
c) surface treatment of the nail, e.g. rough (Fig. 15.1(b)) or galvanised etc.,

Key to Tables 15.1-15.5

Materials
A - aluminium alloy   B - brass   BR - bronze
C - copper   P - plastics   S - steel
SS - stainless steel

Finish/treatment
B - brass   BR - bronze   CP - chromium
G - galvanised   J - jappanned (black)   N - nickel
SC - self-coloured   SH - sherardised   Z - zinc
BZ - bright zinc

Head shape
CKS - countersunk   DM - dome   RND - round head
RSD - raised head   SQ - square

Drive mechanism
SD - Superdriv (Pozidriver)   SL - slotted (screwdriver)
SP - square head (spanner)
(Note: *Superdriv* is the successor to *Pozidriv*.)

d) nail design, i.e. improved nails (Fig. 15.1(c)),
e) dovetail nailing (Fig. 15.1(d)).

Nails are more commonly associated with joints which require lateral support – preventing one piece of timber sliding on another. Nails in this instance are providing lateral resistance (see Fig. 15.2), and, for this to be sustained, resistance to withdrawal is vital.

When nailing wood, splitting can be a problem and can occur when

a) nailing too near to the edge or end of a piece of timber,
b) the nail gauge is too large for the wood section (especially small sections of hardwood),
c) nailing one nail behind another in-line with the grain,
d) using an oversized nail punch,
e) trying to straighten bent nails with a hammer.

If the above cannot be resolved by using other types or sizes or repositioning, then the following remedial measures could be considered;

a) remove the nail's point – N.B. this reduces holding power,
b) pre-bore the timber being fixed,
c) use oval nails,
d) use lost-head nails,

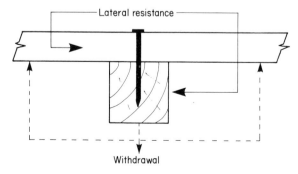

Fig 15.2 Nails - lateral resistance

e) remove bent nails – bent nails never follow a true course.

Figure 15.3 shows examples of minimum nail spacing when fixing to softwood – timber to timber.

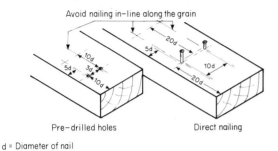

Pre–drilled holes   Direct nailing

d = Diameter of nail

Fig 15.3 Guide for minimum nail spacing when fixing timber to timber - for precise details refer to BS 5268 Part 2. Note: distances refer to space left, not nail centres

15.2 Wood screws

Wood screws have a dual function – not only do they hold joints or articles together; they also act as a permanent cramp, which in most cases can be removed later for either adjustment or modification purposes.

There is a vast variety of screws on the market, and knowing the correct type, size, or shape to suit a specific purpose will become a valuable asset to the joiner. Table 15.2 illustrates several wood-screw fixing devices, together with their use and driving methods. Screw cups and domes (used to conceal yet still provide access to the screw head) are shown in Table 15.3.

Wood-screw labels serve as a quick method of identification, combining an abbreviated description with a screw silhouette or head style and a colour code which represents its base metal (except orange which signifies 'Twinfast' screws). Figure 15.4 shows a typical example of a screw label.

The traditional wood screw has seen several changes over recent years, probably the most significant of these has been the introduction of the *Twinfast* thread which halves the number of turns required to drive the screw home. Then came the *Supafast* threaded screw (see Fig. 15.5), which because of its sharp point and steeply pitched thread not only gives easy starting but drives even

Table 15.2 Wood-screw fixing devices

Screw type	Material	Finish/treatment	Head shape/style	Drive mechanism	Application
Wood screw (Fig. 15.6)	S, B, BR, A, SS	B, BR, CP, J, N, SC, SH, Z	CKS RND RSD	SL SD	Wood to wood; metal to wood, e.g. ironmongery, hinges, locks, etc.
Twinfast wood screw	S, SS	B, SC, SH, Z		SD only	Low-density material; particle board, fibreboard, etc. Drive quicker than conventional screws, having an extra thread per pitch for each turn.
Supafast (Supascrew and Mastascrew) (Fig. 15.5)	S	BZ	CKS	SL, SD	All hard and soft woods as well as man-made boards. Spaced thread and sharp point for quicker insertion. Hardened head and body, so virtually abuse-proof.
Coach screw	S	SC, Z	SQ	SP	Wood to wood; metal to wood (Extra-strong fixing)
Clutch screw	S	SC			Non-removable - ideal as a security fixing
Mirror screw	S, B	CP	CKS DM		Thin sheet material to wood - mirrors, glass, plastics
Dowel screw (double-ended)	S	SC		○	Wood to wood - concealed fastener, cupboard handle, etc.
Hooks and eyes	S	B, CP, SC			Hanging - fixing wire, chain, etc.

Table 15.3 Screw cups and caps

	Screw cups	Cover domes
Material	B, SS	P
Finish/colour	SC, N	Black, white, brown
Shape	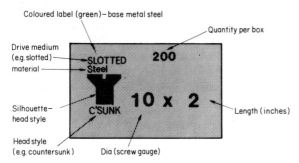 CKS · CKS	DM · DM · DM · SD (a) · (b) · (c)
Application	Countersunk flange increases screw-head bearing area. Used where screw may be re-drawn, e.g. glass beads, access panels, etc.	(a) slots over screw; (b) slots into screw hole; (c) slots into Superdriv screw head. Neat finish yet still indicates location.

Coloured label (green) – base metal steel

Quantity per box

Drive medium (e.g. slotted)

material

SLOTTED
Steel

200

Silhouette – head style

C'SUNK

10 x 2

Length (inches)

Head style (e.g. countersunk)

Dia (screw gauge)

Fig 15.4 Wood screw label identification

faster – its shank diameter is less than the diameter of the thread and therefore splitting the wood is reduced to a minimum. It is particularly useful for screwing into chipboard. Mastascrews and Supascrews which have the supafast thread are hardened after manufacture. This means it is virtually impossible to damage the screws when putting them in or taking them out because the head is as tough as the screwdriver blade or bit – particularly helpful for power driving.

Figure 15.6 shows how materials should be prepared to receive screws, namely by boring

a) a clearance hole to suit the screw shank,
b) a pilot hole for the screw thread,
c) a countersink to receive the screw head – if required.

A bradawl can be used to bore pilot holes in softwood. Failure to use pilot holes could result in the base material splitting and/or losing holding power.

Alternatively, the clearance hole, pilot hole and coutersink or counterbore can be carried out in one operation (see Fig. 2.62 and 2.63). Figure 15.7 and Table 15.4 show and indicate the recommended spacing of wood screws. By using these figures the risk of splitting the base material is reduced and maximum holding power is encouraged.

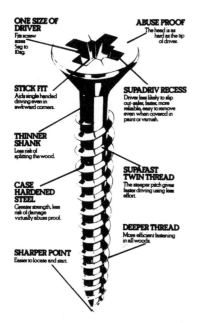

Fig 15.5(a) Nettlefolds Supascrew

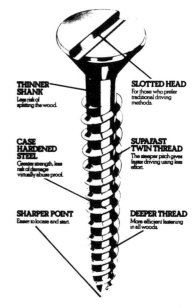

Fig 15.5(b) Nettlefolds Mastascrew

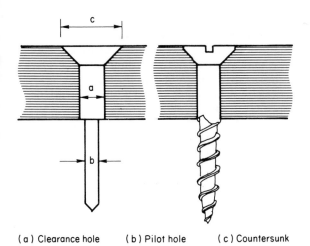

(a) Clearance hole   (b) Pilot hole   (c) Countersunk

Fig 15.6   Preparing material to receive a wood screw

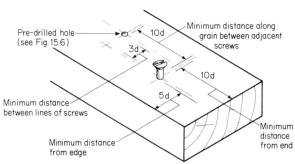

d = Diameter of wood screw shank

Fig 15.7   Minimum screw spacing when driven into pre-drilled holes. Note: when driven without pre-drilled holes minimum distances are, in the majority of cases, much greater - see Table 15.4

Table 15.4   Spacing of woodscrews

	Driven without Pre-drilled Holes	Driven into Pre-drilled Holes
Distance from end	20 D	10 D
Distance from edge	5 D	5 D
Distance between lines of screws	10 D	3 D
Distance along the grain between adjacent screws	20 D	10 D

D = Diameter of Woodscrew.

15.3 Threaded bolts

For the purpose of quick reference, Table 15.5 illustrates those bolts which the carpenter and joiner is likely to encounter. Probably the most common of these is the coach or carriage bolt, and Figs 15.8(a) and (b) show examples of its use. Figure 15.8(c) shows the application of a handrail bolt. The strong hexagonal-headed bolt, together with washers and timber connectors, will be dealt with in connection with roof trusses in *Carpentry and Joinery 2*.

15.4 Fixing plates

Figure 15.9 shows a few of the many fixing plates available. Some plates are multi-purpose, whereas others carry out specific functions; for example, movement plates have elongated slots to allow for

Table 15.5   Threaded bolts

Bolt type	Material	Head	Nut and bolt	Application
Hexagonal head	S			Timber to timber; steel to timber (timber connectors etc.)
Coach bolt (carriage bolt)	S			Timber to timber; steel to timber (sectional timber buildings, gate hinges, etc.)
Roofing bolt	S, A			Metal to metal
Gutter bolt	S, A			Metal to metal
Handrail bolt	S			Timber in its length (staircase handrail, bay-window sill, etc.)

either timber movement and/or fixing adjustment, glass plates act as a hanging medium for fixing items to walls, etc. Multi-purpose plates include angles, straights, tees, etc. to aid or reinforce various joints used in carcase construction.

There are also many different forms of metal straps, framing anchors, and joist hangers etc. which will be dealt with under the heading appropriate to their use, e.g. joist hangers are dealt with in connection with upper floors in *Carpentry and Joinery 2*.

15.5 Plugs

Plugs are used where fixing directly to the base material is impracticable, for example because it is too hard, brittle, or weak. The type of plug or device used will depend on –

a) the required strength of fixing;
b) the type, condition, and density of the base material to receive the plug;
c) whether fixing to a solid or a hollow construction.

A plug is either made from wood or purpose-made from a fibre, plastics, or metal material. The plug hole is bored either by hand, using a plugging chisel and hammer, or by machine, using a rotary or impact drill (depending on the base material – see Chapter 4) and a tungsten-carbide-tipped bit to suit the plug.

Wood plugs   Figure 15.10 shows the shape, preparation, and fixing of a self-tightening plug. This method of plugging should not be used in situations such as those shown in Fig. 15.11 (i.e. at a corner or head) because of the danger of wall disturbance.

Plastics plugs   There is a selection of plastics plugs on the market to suit various base materials. These plastics materials, which include nylon and polythene, have very good holding characteristics and are unaffected by normal temperature change or corrosive conditions in the atmosphere. Figure 15.12 shows the use of a plastics Rawlplug, and its application:

1 Using eye protection drill hole to recommended size.

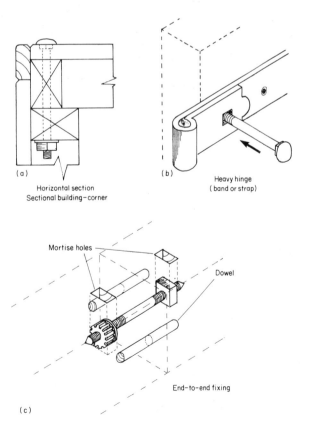

Fig 15.8 Bolt application

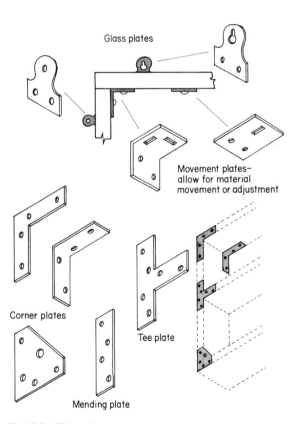

Fig 15.9 Fixing plates

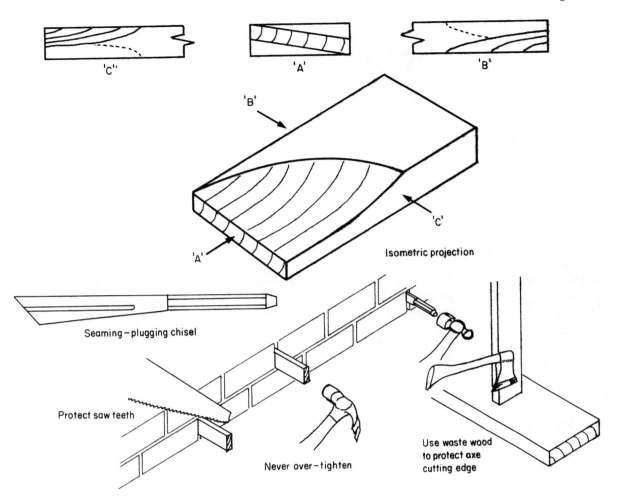

'C'   'A'   'B'

'B'

'C'

'A'

Isometric projection

Seaming – plugging chisel

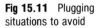

Protect saw teeth

Never over–tighten

Use waste wood
to protect axe
cutting edge

Fig 15.10   Preparing and fixing wood plugs

Fig 15.11   Plugging
situations to avoid

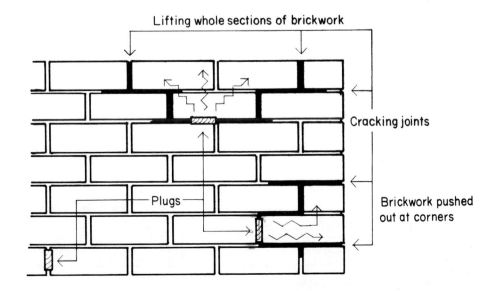

Lifting whole sections of brickwork

Cracking joints

Plugs

Brickwork pushed
out at corners

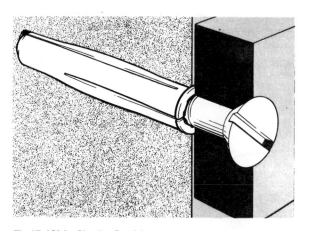

Fig 15.12(a) Plastics Rawlplug

2 Remove boredust by using a screw thread or nail head – *never blow into the hole*.
3 Insert plug.
4 Pass screw through the workpiece and into the plug.
5 Tighten the screw.

By using a plastics plug and sleeve combined, the hole for the fixing can be bored through the workpiece at the same time. This will ensure an accurate final positioning of the workpiece. There are several devices available which allow this practice to be carried out; some use what appear to be the standard type of wood screw, others use a

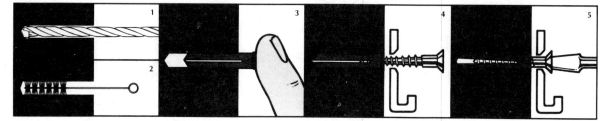

Fig 15.12(b) Method of fixing a Rawlplug

screwnail. Figure 15.13 is an example of a Rawlplug 'Hammer-in fixing' together with its application. Even though these fixings are hammered into place, they may be removed at a later date with a screwdriver.

Cavity fixings These are useful for fixing to hollow or thin materials which are accesible from one side only. Typical situations include

a) hollow walls (building blocks),
b) hollow partitions (plasterboard),
c) wall panelling (plywood),
d) cellular flush doors (hardboard or plywood).

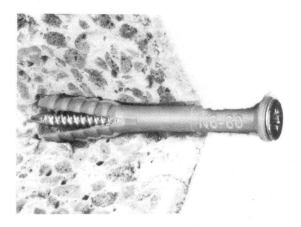

Fig 15.13(a) Rawlplug hammer-in fixing

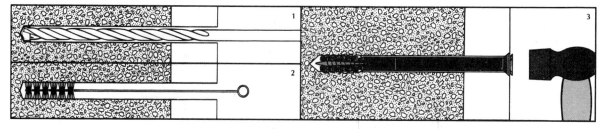

Fig 15.13(b) Method of fixing a Rawlplug hammer-in fixing

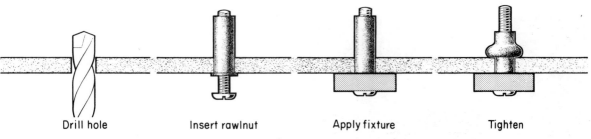

Drill hole   Insert rawlnut   Apply fixture   Tighten

Fig 15.14(b)   Method of fixing a Rawlnut

Fig 15.14(a)   Rawlnut

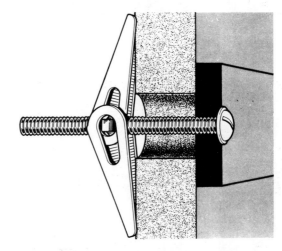

Fig 15.15(a)   Rawlplug spring toggle

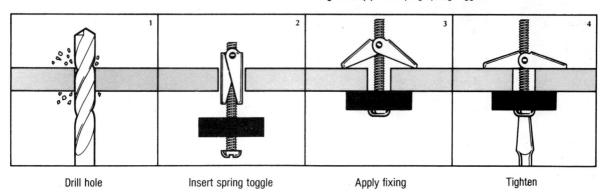

Drill hole   Insert spring toggle   Apply fixing   Tighten

Fig 15.15(b)   Method of fixing a Rawlplug spring toggle

Rawlnut (Fig. 15.14)   This is a flanged rubber sleeve which houses a nut and bolt. As the screw is tightened, the sleeve compresses against the back of the base material.

Spring toggle (Fig. 15.15)   This consists of a screw attached to steel spring wings which when folded back can be pushed through a hole into the cavity, where they spring apart and can then be drawn back against the base material by tightening the screw.

Gravity toggle (Rawlplug Kap toggle) (Fig. 15.16)   This is a metal channel fixed off-centre to two plastic straps and a washer which allows it to swivel. When passed through a hole into a cavity, the channel hangs vertically and on tightening the screw the fixing becomes firm. The fixing is held in position prior to applying the workpiece by the straps which are snapped off once the washer is held firm up against the base material.

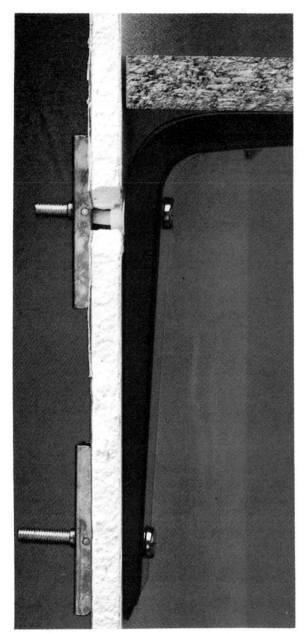

Fig 15.16(a)   Rawlplug Kap toggle

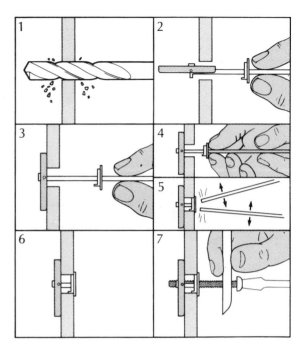

Fig 15.16(b)   Method of fixing a Rawlplug Kap toggle

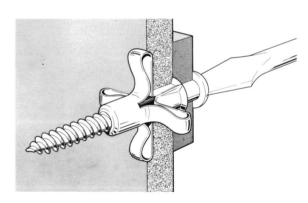

Fig 15.17(a)   Rawlanchor

Fig 15.17(b)   Method of fixing a Rawlanchor

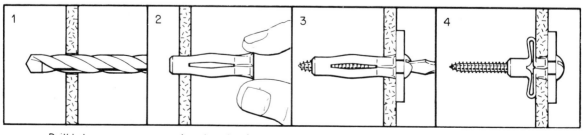

Drill hole   Insert rawlanchor   Apply fixture   Tighten

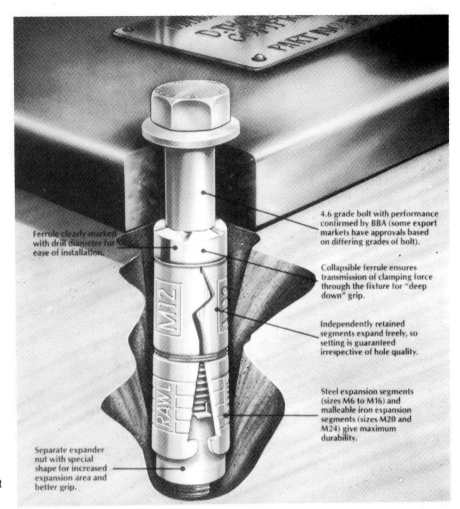

Ferrule clearly marked
with drill diameter for
ease of installation.

4.6 grade bolt with performance
confirmed by BBA (some export
markets have approvals based
on differing grades of bolt).

Collapsible ferrule ensures
transmission of clamping force
through the fixture for "deep
down" grip.

Independently retained
segments expand freely, so
setting is guaranteed
irrespective of hole quality.

Steel expansion segments
(sizes M6 to M16) and
malleable iron expansion
segments (sizes M20 and
M24) give maximum
durability.

Separate expander
nut with special
shape for increased
expansion area and
better grip.

Fig 15.18   Rawlbolt

Rawlanchor Light duty cavity fixing
(Fig. 15.17)   A nylon plug is inserted into an
8 mm hole and, as the fixing screw is tightened,
the plug compresses against the inside face of the
cavity. This is an ideal device for shallow cavities.

Heavy fixings   These include bolts which expand
on tightening while being held in an anchor hole
(usually drilled in concrete). These devices are
generally reserved for heavy structural work. The
most popular of these are *Rawlbolts*, there are several
types and many sizes to suit most heavy duty
situations. Figure 15.18 shows a detailed example of
the components of a Rawlbolt with detachable bolt.
Examples of its use include, anchoring down
machinery and fixing structural components. Figure
15.19 shows a Rawlbolt with bolt projecting,
together with the method of fixing. Examples of use
include securing wallplates and structural
components.

Fig 15.19(a)   Rawlbolt bolt projecting

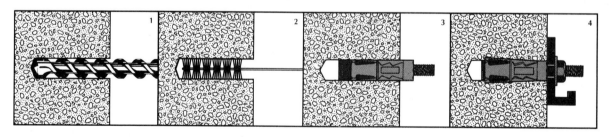

Fig 15.19(b)   Method of fixing a Rawlbolt bolt projecting

Index

Abrasive
    backing material 59
    grains 59
    grits 59
    sheets 59–61
Abrasive wheels 108–10
Adhesives 24–6
Air brick 120–21
Air drying (seasoning) 13–14
Aluminium oxide (abrasive sheet) 59–60
Amperage 94
Angle joints (framing joints) 111, 113–19
Annual rings (growth ring) 2–3
Arch centres 134–40
    applied geometry 135–36
Architrave 118–19
Artificial drying, *see* kiln drying
Auger bits 48–9, 51
Axe 53–4, 169

Back iron, *see* cap iron
Badger plane 39–40
Band-mill 5–7
Band resaw 8
Band-saws 5–7
    narrow-blade band-saw 104–6
Barge board 127–8
Bark 3
    encased 12
Bast (phloem) 3
Batternboard 20
Battens 145, 148–49, 151
Bead 145
Belt sander 92–3, 107
Bench 45, 62–3
Bench holdfast 62–3
Bench hook 36, 62–3
Bench planes 39–46, 39–41, 43–46
Bench rebate plane 39–40
Bench stop 44, 63
Bench vice 35–6, 63
Bevel (sliding) 33–4
Birdsmouth 129, 131
Bits 47–51
    application 49–51
    depth gauge 51
    sharpening 73

    type 47–50
    use 48
Blockboard 20
    sizes 24
Block plane 39–41, 72
Blue-stain (blueing) fungi 16
Bolt
    coach/carriage 167–8
    handrail 167–8
    tower 151
Bond pressure 25
Boring tools 47–51
Brace
    carpenter's 47–51
    ratchet 47–8, 50
    wheel 51
Bracket
    angle 168
    shelf 161–2
Bradawl 51
Bridle joints 116–7
British Standards Institute
    power tools 94
    timber sizes 10–11
Building Regulations
    suspended timber ground floors 121
Bullnose plane 39, 42
Butt hinge 155
Butt joint 111–13

Cabinet scraper 58–61
Cambium 3
Capillarity 156–59
Cap iron 43, 71
Carpenter's brace 47–51
Carriage (coach) bolt 167–8
Carriage plane 39–40
Casein 24
Casement *see* sash
Casement window fastener 155
Casement stay 155
Casement window 152–56
Centre bit 48–9, 51
Centres for arches 134–40
    construction 136–40
    equilateral 137
    segmental 137

semicircular 137–39
Chain mortiser 103, 105
Check to doors 147
Cheek, tenon 115
Chipboard, *see* particle board
Chipbreaker 104–5
Chisels
  grinding 72, 109
  sharpening 71–3
  types 52–3
Chlorophyll 2
Cill (sill) window 152–54, 156, 159
Circular plane (compass plane) 39, 41
Circular saw
  bench 98–100
  mouthpiece 99
  packing 99
Circular sawing machines 5–6, 98–101
Claw hammer 55–6
Club (lump) hammer 55
Coach bolt (carriage bolt) 167–8
Coach screw 165
Cogged joints 117
Collar roof 130–1
Combination gauge 33
Combination plane 39
Combination square 29–30, 32
Common rafter 128–32
  setting out 131
Compass plane (circular plane) 39, 41
Compression failure (upset) 11–12
Concrete
  cast in-situ 141–2, 144
  pre-cast 141–2
Conifers (coniferous trees) 3–4
Contact adhesive 24–6
Conversion timber 5–9
  geometry 9
  methods 6–9
Coping saw 37
  blades 37
  use 37
Countersink
  bits 48–9
  use 48–51
Couple roof 131
Cramp
  application 61–2
  types 61–2
Cross-cut handsaws 34–5, 37
Cross-cutting machines 97–8
Crown
  of floor joists 122
  tree 1
Crown guard 99–101
Cupping 16, 112
Cup shake 11
Cutter block 101–3
Cutting gauge 33
Cutting list 29, 30, 77, 79, 84

Damp proof course (d.p.c) 120–1
Dead knots 13
Deciduous trees 3–5
Degraded timber 11–13
Depth gauge, drill 51
Dermititis 26
Diamonding 16
Dimension saw bench 99–100
Disc sander 107
Dividers 29–30, 119
Doors, *see* matchboarded doors, ledged-and-braced-
    battened doors
Double insulation 94–5
Double roof 127
Double tenon 115
Dovetail
  halving 114
  housing 114
  joint 117–9
  lapped 117
  nailing 164
  proportions 117
  saw 38–9
  template 34, 118
Dowel 113
Dowel bit 48–9
Dowel cradle 63
Dowel joints 112–13, 116
  preparation 113
Dowel screw 165
Draw-boring 148, 150
Drill
  bits 48
  electric 88–91
  attachments 89–90
  speeds 88–89
  stand 90
Drying timber 13–16
Durability, timber 17
Dust extraction 107–8, 110

Earlywood (springwood) 2–3
Earthing 94–5
Easing and striking
  centres 104
  formwork 141
Eaves 132–3
Electric drill 88–91
  attachments 89–90
Electric tools, *see* portable powered hand tools
Equilateral arch 137
Evergreen trees 3–4
Expansive (expansion) bit 48–9
Exterior-grade plywood 19

Face side and edge 30–2
Fascia board 127–8
Felling trees 1
Fibre board 21–2
Figure (grain) 9

Files 66, 68–70, 73
Finger joint 111
Firmer chisel 52
Firmer gouge 52
Fixing devices 163–74
Fixing plates 167–8
Flat bit 48–9
Flatting and edging 43–5, 100
Floor boards
    laying and fixing 123–5
    section 123
Forest farming 1
Formwork 141–4
Forstner bit 48–9
Framed saws 37–8
Frame saw (gang saw) 5–6
Framing joints 111, 113–19
Fruiting bodies (sporophores) 17
Fungi 16–17
Furniture beetle 17–18

Gable end 127–28, 132
Gable-ended roofs 126–32
Gable-ladder 127–8, 131
Gang saw (frame saw) 5–6
Gap filling 25
Garnit paper 59–60
G cramp 61–2
Geometry
    arches 134–36
    roof rafters 131
    timber conversion 9
Glasspaper 59–60
Glues 24–26
Gouges
    in-cannal 52
    out-cannal 52
    sharpening 73
    types 52
Grain sloping 12
Grinding
    chisels 72, 109–10
    plane blades 70–1, 109–10
Grinding machines 108–9
Grinding wheels 109–10
Grindstones, see grinding wheels
Ground floors, suspended timber 120–26
Grounds, timber 26
Growth of a tree 1–3
Growth rate 11–12
Growth rings (annual rings) 2–3

Hacksaws 37–8
Half-lap joints 114
Halving joints 114
Hammers 55–6
    types 55
    use 56
Hand drill 51
Hand-feed planer and surfacer 100–1

Handrail bolt 167–8
Handsaws 34–8
Hardboard 21–22
    sizes 24
Hardwood 3–5
Haunch 115–16, 150
Heart shake 11–12
Heartwood 2–3
Heel and toe
    belt sander 92–3
    plane 44
    saw 69–70
Hinges
    butt 155
    cranked 155
    tee 149, 151
Holdfast, bench 62–3
Holding-down springs 101–2
Hollow-chisel mortiser 103–4
Hollow floors, *see* suspended timber ground floors
Hollow-ground blades 109
Honing 69–73
Housing joints 114
    cutting 114–15
Hygroscopic material 15
Hyphae 16–17

Infeed table 101–2
Insect, wood-boring 17–18
in-situ concrete 141–2
Insulation board 18, 21–2
    sizes 24
Interior-grade plywood 19
Ironmongery
    for matchboard doors 151
    for casement windows 155
    list 30

Jennings bit 48–9
Joints
    for framing 113–19
    for lengthening 111–12
    for widening 112–13
Joist brace (upright brace) 48
Joists 120–23
    laying and levelling 122–3

Kerf 38
Keyhole, cutting 151
Keyhole saw 37
Kiln drying (seasoning) 13–16
Knots 13
Knotting 151

Lagging (centres) 135–7
Laminated plastics 22–4
    application 23
    cutting 22–3
    sizes 24
Laminated timber 111

Laminboard 19–20
    sizes 24
Lapped dovetail 117
Larvae 18
Latch, Suffolk and Norfolk 151
Latewood (summerwood) 3
Lathe 105–6
Leaves 1–2
Ledged-and-braced battened doors, *see* matchboard
    doors
Locks 151
Log 5
Lump hammer 55

Machine-shop layout 110
Machines, woodworking 97–110
Maintenance of tools 66–73
Mallet 56–7
Manufactured boards 18–22
    sizes 24
Marking gauge 30, 33
Marking-out tools 29–34
Matchboard 145
Matchboard doors 145–51
    assembly 147–49
    construction 151
    fitting and hanging 149–50
    frame 150
    hardware (ironmongery) 151
Measuring tools 27–8
Medium board 22
    sizes 24
Medium density fibre board (MDF) 22
Medulla 2–3
Meniscus 157–8
Mitre block 62–3
Mitre box 62–3
Mitre fence 99
Mitre joint 118–19
Mitre square 33
Mitre template 33
Moisture content 13–16
Moisture movement 15, 157–9
Moisture-resistance (MR) adhesive 25
Monopitch roof 131
Mortise chisel 52–3
Mortise gauge 33
Mortise hole
    hand-cut 52–3, 116
    machine-cut 104
Mortise and tenon 114–16
Mortiser
    chain type 103–5
    hollow chisel type 89, 91, 103–4
Mortising machines 89, 91, 103–5
Mould box 141, 143–4
Mycelium 16–17

Nail punch 147–8
Nails 163–5

Narrow-blade band-saw 104–6
Natural defects 11–13
Natural drying (air drying) 14
Norfolk latch 150–1
Notched joints 117

Oilstone 66–7
    box 67
    slip, *see* slipstone (67)
    use 69, 71–3
Opening light 154–5
Orbital sander 93
Outfeed table 101–2

Packing, circular saw 99
Panal board, see medium board
Panel planer 102–3
Panel saw 35, 37
Panel sawbench (machine) 100–1
Paring 52–3
Paring chisel 52
Particle board 20–2
    sizes 24
Phenol formaldehyde (MF) 24–5
Phloem (bast) 3
Photosynthesis 2
Piercing (narrow blade saw) 37
Pith (medulla) 2–3
Plane irons 43
Planes 39–47
    bench 39–41
    special-purposes 41–44
Planing
    by hand 44–47
    by machine 10–12, 88, 100–3
Plastic laminates, laminated plastics 22–3
    sizes 24
Plastic plugs 168, 170
Plastics screw domes 166
Plates, fixing 167–8
Plough plane 39, 42
    application 46
Plugging (seaming) chisel 168–9
Plugs
    plastics 168, 170
    wood 126, 168–9
Plumb stick, (*see also* spirit level)
Plywood 19–20
    sizes 24
Polyvinyl acetate (PVA) 24–5
Portable powered hand tools (electric) 88–96
    electricity supply 93–5
    safety 96
    specification plate 93–4
Porterbox storage system 74–87
    drop-fronted tool box & saw stool 74–8
    Portercaddy 78–81
    Portercase 79, 81–2
    Porterchest 83–7
    Porterdolly 82, 79

Posidriv screws 165–7
   screwdriver 57–8
Pot life (adhesives) 25
Power 93
Pre-cast concrete 141, 143–4
Preservatives 17–18, 121
Press lock 151
Pressure bars 101–2
Priming paint 151
Propping (centres) 140
Purlin roof 127

Quadrant 9
Quarter-sawn boards 8–9
Quartered log 9

Radial-arm cross-cut saw 98
Radial-sawn boards 9
Radial shrinkage 16
Radius 8, 134–6
Rafter, *see* common rafter
Rate of growth 11–12
Rays 3, 9
Rebate
   cutting 46, 113
   for doors 149
   for windows 154
Rebate planes 39, 42–3, 46
Reduced voltage 95
Regulations
   abrasive wheels 109
   protection of eyes 110
   timber ground floors 121
   woodworking machines 107–8
Resorcinol formaldehyde (RF) 24–5
Ribs 136
Ridge 127–8
Ridge board 127–8
Rift-sawn boards 8–9
Rim dead lock 151
Ripping
   by hand 34–6
   by machine 98
Riving knife 99
Roof
   assembly 129, 131–2
   geometry 131
   model 130
   pitch 127–8
   rafters, *see* common rafters
   terminology 127
   types 127, 131
Rotary impact drill 88, 90–1
Router 39
   application 42, 114
Rule
   flexible tape 27–8
   four-fold 27–8
   scale 27

Safety (*see also* regulations)
   portable powered hand tools 96
   precautions (adhesives) 26
Sanders
   belt 92–3, 107
   orbital 93
Sanding
   abrasives 59–61
   blocks 61
   machines 92–3, 107
   operations (hand) 61
Sap 1–2
Sap-stain fungi 16
Sapwood 2
Sash
   opening 154
   rail 152–4
   stile 152–4
Sash cramp 61–2
Saw (*see also* circular saws)
   blade protection 65–6
   blades 37
   cut (kerf) 38
   gullet 38
   handles 37
   set 38, 68
   setting 68–9
   shaping 68–9
   sharpening 69–70
   teeth 37–8
   topping 68
Sawing 35–8
Sawmills 5–9
Sawn sections sizes 10–11
Saw stool 62–3
Saw vice 62–3
Scale rule 27
Scarf joint 111
Scrapers
   chisels 106
   hand-held 58–61
Screwdriver
   efficiency 58
   electric 91–2
   maintenance 73
   ratchet 57
   rigid-blade 57
   spiral (pump action) ratchet 57–8
Screws 165–67
Scribing
   joints 118–19
   methods 29, 118–19
Scribing gouge 52
Seaming chisel 168–9
Seasoning *see* drying timber
Second fixing 125
Segmental arch 135
Setting-out
   arches 134–9
   doors 29–31

rafters 131
rods 29–31
tools 29–34
Shakes 11–12
Sharpening
    bits 73
    chisels 72–3
    plane blades 68–72
    saws 68–70
Shelf
    brackets 160–2
    support 160–2
    types 161
Shelf life (adhesives) 25
Shelving 160–2
Shoulder, tenon 115
Shoulder plane 39, 42
Shrinkage 15–16
Shuttering, *see* formwork
Side rebate plane 39, 42
Sill (cill), window 152, 154, 156
Single roof 127
Sizes
    boards 24
    timber 10–11
Skirting board 125–6
    fixing 126
    patterns 125
Slab 7–8
Sliding bevel 33–4
Slipstone 66–7
    box 67
    use 73
Sloping grain 12
Slot-screwed joint 112–13
Soffit 127–8, 132–3
Softboard, *see* insulation board
Softwood 3–4
Spar, *see* common rafter 127–9
Special-purpose planes 39, 41–7
Specification plate 93–4
Spirit level 122, 150
Spokeshave 39, 47
    blade holder 68
Sporophores 17
Spinging line 136
Springwood 3
Square
    combination 30, 32
    try 30, 32
    testing 32
Staples 163
Star dowel 163
Star shake 11–12
Stem (trunk) 1–2
Stopped housing 114
Storage life (adhesives) 25
Straight-edge 122
Stropstick 66, 71
Structural defects 11–13

Stub tenon 115
Suffolk latch 150–1
Summerwood (latewood) 3
Superdriv screws 165–6
Surface tension 157–9
Surfacing machines 100–1
Surform tools 54
Suspended timber ground floors 120–6
Synthetic resins, *see* adhesives

Tangent 9
Tangential-sawn boards 8–9
T bar cramp 61–2
Tee hinge 151
    fixing 149–50
    types 151
Tempered hardboard 22
Tenon
    cheek 115
    mortice and 114–16
    shoulder 115
Thumb latch 150–1
Thicknesser 102–3
Throating 158–9
Through-and-through sawing 7–8
Tie, *see* centres for arches and roof
Tilting arbor 100
Tilting table 105, 107
Timber
    cladding 112
    conversion 5–9
    defects 11–13, 16
    drying (seasoning) 13–16
    durability 17
    figuring 9
    floors 120–26
    hardwoods 3–5
    nomenclature 3–5
    preservation 17–18, 121
    quantity 10
    seasoning, see drying
    sections 10–12
    shakes 11–12
    shrinkage 15–16
    size and selection 10–12
    softwood 3–4
    sources 4–5
    structural defects 11–13
    supply 5
Toe, *see* heel and toe
Tongue and groove 112–13, 123–5, 145
Tongue, loose 112
Tool box (Porterbox storage system) 74–87
Tools
    boring 47–51
    chisels 52–3
    driving 55–8
    finishes 58–61
    holding 61–3, 77
    maintenance 66–73

marking-out 29–33
measuring 27–8
planing 39–47
sawing 34–8
setting-out 29–33
shaping 53–4
storage 62–6
Topping (saw) 68–9
Tower bolt 151
Trammel 30
Tree
  crown 1
  growth 1–3
  roots 1–2
  stem (trunk) 1–2
Trenching, *see* housing joint
Trestle 62–3
Trim 112
  architrave 118–19
Try plane 39–40, 44–5
Try square 30, 32
  testing 32
Turning piece 134
Twin tenon 115

Upset (compression failure) 11–12
Urea formaldehyde 24–5

V (vee) joint 112, 151
Veneers
  application 23
  cutting 19

plastics, see laminated plastics
Ventilation of subfloor space 121
Vice
  bench 62–3
  saw 66, 68
Voltage
  reduction 94–5

Wallplate 120–1, 123, 128–9, 131–2
Wane (waney edges) 12
Warrington hammer 55–6
Wattage 94
Wheel-brace 51
Window
  board 156
  casement 152–6
  ironmongery 155
  sill (cill) 152, 154, 156
Wood (*see also* timber)
  earlywood (springwood) 3
  defects 11–13
  latewood (summerwood) 3
  species 4–5
  springwood (earlywood) 3
  summerwood (latewood) 3
Wood-boring beetles 17–18
Wood screws 165–7
Wood-turning lathe 106
Wood turning tools 106
Wood working machines 97–110
Woodworm, *see* wood-boring beetles